Springer Monographs in Mathematics

This series publishes advanced monographs giving well-written presentations of the "state-of-the-art" in fields of mathematical research that have acquired the maturity needed for such a treatment. They are sufficiently self-contained to be accessible to more than just the intimate specialists of the subject, and sufficiently comprehensive to remain valuable references for many years. Besides the current state of knowledge in its field, an SMM volume should ideally describe its relevance to and interaction with neighbouring fields of mathematics, and give pointers to future directions of research.

Heiko Dietrich · Bettina Eick

Coclass graphs of p-groups

Springer

Heiko Dietrich
School of Mathematics
Monash University
Clayton, VIC, Australia

Bettina Eick
Institut Analysis und Algebra
Technische Universität Braunschweig
Braunschweig, Germany

ISSN 1439-7382 ISSN 2196-9922 (electronic)
Springer Monographs in Mathematics
ISBN 978-3-032-17505-2 ISBN 978-3-032-17506-9 (eBook)
https://doi.org/10.1007/978-3-032-17506-9

Mathematics Subject Classification: 20D15

This Springer imprint is published by the registered company Springer Nature Switzerland AG
The registered company address is: Gewerbestrasse 11, 6330 Cham, Switzerland

Preface

A **finite p-group** is a group whose order is a power of a prime p. Sylow's theorem (1872) states that every group whose order is divisible by a prime power p^n contains a subgroup of that order. This fundamental result profoundly influenced group theory and became a cornerstone for understanding the significance of finite p-groups.

Finite p-groups exhibit a rich and well-developed structure theory. A key result in this area states that every non-trivial finite p-group has a non-trivial centre, implying that such groups are nilpotent. Moreover, every maximal subgroup is normal and has index p. The books of Berkovich and Berkovich & Janko (2008 – 2018) provide a wealth of information on the structure of finite p-groups.

The theory of finite p-groups also plays a crucial role in the broader study of finite groups. Interesting results about finite groups can be derived from their p-subgroups. For instance, the Schur multiplier of a finite group is determined by its p-components, with each p-component isomorphic to a subgroup of the Schur multiplier of a Sylow p-subgroup. Furthermore, in modular representation theory, some prominent conjectures and theorems rely on a detailed understanding of the p-local structures of finite simple groups, specifically, a classification of their p-subgroups along with their centralisers and normalisers. These and many other results underscore the fundamental importance of finite p-groups in the theory of finite groups.

Enumerations and asymptotics. The investigation of the number $\mathrm{gnu}(m)$ of isomorphism types of finite groups of order m is an active research area in finite group theory, see for example Conway, Dietrich & O'Brien (2008) or the book of Blackburn, Neumann & Venkataraman (2007). For instance, if $\mu(m)$ denotes the largest exponent of a prime in the prime power factorisation of m, then Pyber (1993) showed that

$$\mathrm{gnu}(m) \leqslant m^{\frac{2}{27}\mu(m)^2+O(\mu(m)^{5/3})}.$$

The dominance of p-groups among all finite groups is highlighted by comparing this estimate with the formula for prime powers that resulted from work of Higman (1960) and Sims (1965), with improvements by Newman & Seeley: if $m = p^n$, then

$$\mathrm{gnu}(m) = p^{\frac{2}{27}n^3+O(n^{5/2})} = m^{\frac{2}{27}\mu(m)^2+O(\mu(m)^{3/2})}.$$

Explicit calculations by Besche, Eick & O'Brien (2002) (with a correction for order 1024 by Burrell (2021)) support the dominance of p-groups also for small orders: it is shown that 49497921211 of the 49910531351 groups of order at most 2000 have prime power order (not equal to 1), and 49487367289 of these have order 2^{10}.

Higman (1960) considered the function $\mathrm{gnu}(p^n)$ for fixed n as a function in p and introduced his famous PORC Conjecture: This states that for fixed n, the number $\mathrm{gnu}(p^n)$, as a function in p, is a polynomial on residue classes. The PORC Conjecture was proved for $n \leqslant 7$ by O'Brien & Vaughan-Lee (2005), and remains open beyond this.

Classifications and algorithms. Every finite p-group can be defined via a power-commutator presentation, a structural description well-suited for computation. A practical algorithm to determine such a presentation is the p-quotient algorithm developed in a sequence of papers by Macdonald (1974), Havas & Newman (1980), and Newman & O'Brien (1996). Given a finitely presented group G, this algorithm determines a power-commutator presentation for the largest p-quotient of G if it exists. A variation of this algorithm has been used by O'Brien & Vaughan-Lee (2002) in the construction of power-commutator presentations for various restricted Burnside groups. For example, the restricted Burnside group $R(2,7)$ has order 7^{20416}, which illustrates the power of this algorithm.

A related algorithm is the p-group generation algorithm by Newman (1977) and O'Brien (1990). In principle, it is able to determine a complete and irredundant set of isomorphism type representatives of the finite p-groups of a given order p^n. This algorithm was used to classify the groups of order dividing 2^9, 3^8, and p^7 for many small primes p. Similar techniques have also been applied to enumerate the groups of order 2^{10}.

A finite p-group of order p^n with $n \geqslant 2$ has (nilpotency) class at most $n-1$. It has **maximal class** if its class is precisely $n-1$. First examples of classifications of p-groups of maximal class were already provided by Blackburn (1958) who determined the finite 2- and 3-groups of maximal class. There are infinitely many 2- and 3-groups of maximal class, but Blackburn described them by a finite set of parametrised group presentations.

The finite p-groups of maximal class have been studied extensively in the literature. Wiman (1952) investigated their structure and Leedham-Green & McKay (1976 – 1984) exhibited highly interesting insights into the graphs associated with these groups. Moreover, Fernández-Alcober (1995), Miech (1982), and Vera-López et al. (2011) have studied in detail the commutator structure of the p-groups of maximal class.

Coclass theory. A finite p-group of order p^n and class c has coclass $n-c$. Thus, coclass generalises the concept of maximal class. Leedham-Green & Newman (1980) suggested the investigation of finite p-groups by using the **coclass** as a primary invariant. They provided five conjectures on the structure of the finite p-groups with fixed coclass. These **Coclass Conjectures** became a central focus in the research area. Building on the work of numerous mathematicians, these conjectures were ultimately proved independently by Shalev (1994) and Leedham-Green (1994). The book by Leedham-Green & McKay (2002) provides an excellent introduction to this area of research, and also proofs of the five conjectures. It is beyond the scope of the present book to repeat these proofs.

In recent years, research in coclass theory has shifted toward investigating **coclass graphs**. The vertices of the coclass graph $\mathcal{G}(p,r)$ correspond to isomorphism type representatives of all finite p-groups of coclass r, and two vertices are connected by an edge if their associated groups satisfy the condition that one is a quotient of the other by a normal subgroup of order p. These graphs have an interesting structure theory and an enormous impact on the classification of finite p-groups by coclass. Explicit computations have revealed surprising periodicity patterns, which led to the following central conjecture in Coclass Theory.

Coclass Graph Conjecture. *Let p be a prime and $r \in \mathbb{N}$. The coclass graph $\mathcal{G}(p,r)$ can be constructed from a finite subgraph using two types of periodic patterns; one suffices for $p = 2$.*

A related conjecture is that these graph theoretic periodicities are reflected in the structures of the underlying groups; specifically, it is conjectured that the infinitely many finite p-groups of coclass r can be described by finitely many parametrised group presentations, similarly to Blackburn's classification of the 2- and 3-groups of coclass 1.

A first breakthrough showed that the Coclass Graph Conjecture holds for all graphs $\mathcal{G}(2,r)$. This was proved independently by du Sautoy (2001) and by Eick & Leedham-Green (2008). The latter proof was constructive and demonstrated that the observed graph theoretic periodicities lead to a classification of the underlying groups: it was proved that the infinitely many groups in $\mathcal{G}(2,r)$ can be described by finitely many group presentations that each contain an integer parameter. The infinitely many groups that are described by one of these parametrised presentations form a *coclass family*. Further applications show that the groups in a coclass family share significant features: for example, their automorphism groups (Eick (2006), Couson (2014)), their Schur multipliers (Eick (2008), Eick & Feichtenschlager (2011), Moravec (2011)), their cohomology (Eick & Green (2017)), and their character degrees (Couson (2014)) can be described simultaneously. This illustrates that coclass families are a powerful tool to investigate the structure of finite 2-groups by coclass.

For odd primes, the structure of the coclass graph $\mathcal{G}(p,r)$ is known to be much more complex than for $p = 2$. Partial results of computational and theoretical nature exist, but a full classification as in the case $p = 2$ is not *yet* reached. However, if the central conjecture is proved one day (and we believe that this is possible), then Coclass Theory has the potential to bring order into the otherwise chaotic realm of finite p-groups!

Other invariants? Coclass has of course not been the only invariant that has been used for classification approaches in p-group theory. Significant efforts have gone into classifying p-groups by order, which is partly driven by the aforementioned PORC conjecture. One could also ask for a classification of the finite p-groups of a fixed class or derived length. However, these approaches share the problem that with increasing class or derived length, new structural phenomena seem to occur. In fact, every finite p-group of class (or derived length) c arises as a direct factor of some finite p-group of class (or derived length) $c + 1$ which illustrates that these invariants are not capable of providing a clear separation between successive stages in this progression.

On this book. The goal of this book is to provide a gentle introduction to coclass theory, to present in a unified framework the key developments of the past two decades, and to discuss open conjectures. Alongside a systematic exposition of known results, we also include new findings and corrections to existing proofs in the literature, with each chapter's introduction highlighting these contributions. A particular emphasis is placed on the interplay between the intricate structure of p-groups of coclass r and their graphical representations in the form of coclass graphs. Our aim is to acquaint readers with the many deep and fascinating results of coclass theory, while also offering a solid foundation and a reliable reference source.

This book is intended for readers with a graduate-level background in group theory and general algebra. Although we have aimed to keep it largely self-contained, some advanced results from p-group theory are cited where necessary. Experienced readers may prefer to bypass the introductory chapters.

Acknowledgements. While we wrote this book, we enjoyed the hospitality of various institutions. In particular, we thank the *Mathematisches Forschungsinstitut Oberwolfach* for accepting us for a research-in-pairs visit. We also thank Charles Leedham-Green and Mike Newman for introducing us to the wonderful topic of coclass theory. Their insights into this topic are invaluable. We are grateful to our friend Eamonn O'Brien who discussed earlier drafts of this book with us and contributed many helpful comments.

Braunschweig, Melbourne
December 2025

Heiko Dietrich
Bettina Eick

Contents

1 Introduction to coclass graphs 1
1.1 Fundamental concepts of group theory 1
1.2 The definition of coclass 6
1.3 Introduction to pro-p-groups 7
1.4 The *Coclass Conjectures* 13
1.5 The definition of the coclass graph $\mathcal{G}(p, r)$ 14
1.6 The examples $\mathcal{G}(2, 1)$ and $\mathcal{G}(3, 1)$ 17

2 Constructive methods and visualisations 21
2.1 Algorithmic methods 21
2.2 Investigating coclass trees 28
2.3 The graph $\mathcal{G}(2, 2)$ 30
2.4 A conjectural description of $\mathcal{G}(5, 1)$ 31
2.5 Towards periodic patterns in coclass graphs 32

3 Number theory and group constructions 35
3.1 From coclass theory to number theory 35
3.2 Introduction to p-adic numbers 36
3.3 The field $\mathbb{Q}_p$ and its finite extensions 40
3.4 Cyclotomic extensions of $\mathbb{Q}_p$ 41

4 Matrix groups and uniseriality 47
4.1 Uniseriality and two-step centralisers 47
4.2 Construction of finite p-groups with uniserial action 50
4.3 Structure of finite p-groups with uniserial action 58
4.4 Space groups with p-adic translation subgroup 64
4.5 Infinite pro-p-groups of finite coclass 66
4.6 Application: The class-breadth conjecture 69

5 Cohomology theory and extensions ... 73
5.1 Basic definitions ... 73
5.2 Sequences in cohomology ... 81
5.3 A cohomology splitting ... 83
5.4 The coclass of an extension ... 88
5.5 The isomorphism problem for space groups ... 92

6 The number of infinite paths in $\mathcal{G}(p, r)$ ... 95
6.1 The number of uniserial p-adic space groups ... 96
6.2 The number of central extensions ... 98
6.3 An upper bound ... 99

7 Periodicity I ... 101
7.1 Pruned branches and extension structures ... 101
7.2 The theorem of Periodicity I ... 104
7.3 Coclass families and parametrised presentations ... 109
7.4 Automorphism groups and coclass families ... 112
7.5 Applications: Schur multipliers, character theory, and more ... 113

8 The skeletons of a coclass tree ... 117
8.1 The exterior square and homomorphisms ... 117
8.2 Skeleton groups in the split case ... 123
8.3 Skeleton groups in the non-split case ... 124
8.4 The isomorphism problem for skeleton groups ... 128
8.5 Constructible groups versus skeleton groups ... 134

9 Groups of maximal class ... 137
9.1 The broad structure of $\mathcal{G}(p, 1)$ ... 137
9.2 Skeleton groups in $\mathcal{G}(p, 1)$ ... 139
9.3 Galois trees in $\mathcal{G}(p, 1)$ ... 145

10 Towards Periodicity II ... 159
10.1 The general case ... 159
10.2 The maximal class case ... 160

References ... 163

Index ... 167

List of Symbols ... 169

Chapter 1
Introduction to coclass graphs

In this chapter we exhibit the fundamental concepts of group theory underpinning the coclass theory of finite p-groups. These concepts will be familiar to most readers and we develop them fairly rapidly. We then introduce the main theme of this book – coclass graphs – and present some easy examples for these. A particular focus is on a gentle introduction to an algebraic treatment of infinite pro-p-groups.

1.1 Fundamental concepts of group theory

This section provides the concepts of group theory and, in particular, of finite p-groups, that are used in later chapters.

1.1.1 Commutators and subgroup series

If g and h are elements of a group G, then the conjugate of g by h is $g^h = h^{-1}gh$ and the commutator of g and h is $[g, h] = g^{-1}h^{-1}gh$. If X and Y are two subsets of G, then $[X, Y]$ is the subgroup generated by all commutators $[x, y]$ with $x \in X$ and $y \in Y$. The *derived subgroup* of G is $G' = [G, G]$; equivalently, G' can be defined as the smallest normal subgroup of G with abelian quotient G/G'. The *centre* of G is the subgroup $Z(G) = \{g \in G \mid gh = hg \text{ for all } h \in G\}$. We write $U \leqslant G$ if U is a subgroup of G, and $U < G$ if U is a proper subgroup.

Definition 1.1 Let G be a group.

a) The *derived series* of G is $G = G^{(0)} \geqslant G^{(1)} \geqslant \ldots$ where $G^{(i+1)} = [G^{(i)}, G^{(i)}]$ for each $i \geqslant 0$.
b) The *lower central series* of G is $G = \gamma_1(G) \geqslant \gamma_2(G) \geqslant \ldots$ where $\gamma_{i+1}(G) = [\gamma_i(G), G]$ for $i \geqslant 1$.
c) The *upper central series* of G is $\{1\} = \zeta_0(G) \leqslant \zeta_1(G) \leqslant \ldots$ where $\zeta_{i+1}(G)$ is defined by $\zeta_{i+1}(G)/\zeta_i(G) = Z(G/\zeta_i(G))$ for each $i \geqslant 0$.

H. Dietrich and B. Eick, *Coclass graphs of p-groups*, Springer Monographs in Mathematics, https://doi.org/10.1007/978-3-032-17506-9_1

A group G is *soluble* if its derived series terminates at the trivial subgroup. It is *nilpotent* if its lower central series terminates at the trivial subgroup, or, equivalently, if its upper central series terminates at G. If G is soluble and ℓ is minimal with $G^{(\ell)} = \{1\}$, then G has *derived length* ℓ. If G is nilpotent, then its lower and upper central series have the same length and this length is the *(nilpotency) class* of G; that is, if c is minimal with $\gamma_{c+1}(G) = \{1\}$, then $\mathrm{cl}(G) = c$ is the class of G.

The terms of the derived series and of the lower central series of a group G are *fully invariant* subgroups, that is, if $\pi\colon G \to H$ is a homomorphism, then $(G^{(i)})^\pi \leqslant H^{(i)}$ and $\gamma_i(G)^\pi \leqslant \gamma_i(H)$ for all i, with equality if π is an epimorphism. This also implies that $G^{(i)}$ and $\gamma_i(G)$ are *characteristic* in G, that is, they are invariant under each automorphism of G. The subgroups of the upper central series are also characteristic, but not necessarily fully invariant.

The product of $A, B \leqslant G$ is the set $AB = \{ab \mid a \in A, b \in B\}$. It is well known that AB is a subgroup of G if and only if $AB = BA$ as sets; the latter holds, for example, if one of A and B is a normal subgroup of G.

Lemma 1.2 *If A and B are nilpotent normal subgroups of a group G, then the product AB is a nilpotent normal subgroup of G.*

Proof Since A and B are normal, the product AB is a normal subgroup of G. A straightforward calculation shows that $[ab, c] = [a, c]^b[b, c]$ for all $a, b, c \in G$, and thus $[UV, W] = [U, W][V, W]$ and $[U, VW] = [U, V][U, W]$ for all $U, V, W \trianglelefteq G$. Hence $\gamma_2(AB) = [AB, AB] = [A, A][A, B][B, B]$ and, by induction, $\gamma_i(AB)$ is a product of iterated commutators of i groups in $\{A, B\}$. By assumption, A and B are nilpotent, say of class n and m, respectively. If $i \geqslant n + m + 1$, then every iterated commutator of i groups in $\{A, B\}$ has at least $n + 1$ copies of A or at least $m + 1$ copies of B. Since $[A, B] \leqslant A \cap B$ due to A and B being normal, this implies that $\gamma_i(AB) = \{1\}$. Thus, AB is nilpotent. □

1.1.2 Groups of prime-power order

Groups of prime-power order (finite p-groups for short) are the central topic of this book. We recall their definition for completeness.

Definition 1.3 A finite group G is a *p-group* if its order $|G|$ is a power of a prime p.

Recall that the index $[G : U]$ of a subgroup U in a group G is the number of right cosets of U in G; if G is finite, then $[G : U] = |G|/|U|$. Lagrange's Theorem asserts that the order of each element and each subgroup in a finite p-group is a power of p, and the same is true for the index of a subgroup and, by the Orbit-Stabiliser Theorem, for the size of a conjugacy class.

The following proposition collects a few important results on finite p-groups. For a group G acting on a set M we denote by $\mathrm{Fix}_G(M)$ the set of all $m \in M$ that are fixed by every $g \in G$.

Proposition 1.4 *Let G be a group of order p^n.*

a) *If G acts on a finite set M, then* $|\mathrm{Fix}_G(M)| \equiv |M| \bmod p$.

b) *If G is non-trivial, then the centre $Z(G)$ is non-trivial.*

c) *If G is non-abelian, then $G/Z(G)$ is not cyclic.*

d) *The group G is nilpotent; if $n > 1$, then the class is less than n.*

e) *The group G has a normal subgroup of order p^i for every $i \in \{0, 1, \ldots, n\}$.*

f) *If $N \trianglelefteq G$ is a non-trivial normal subgroup, then $N \cap Z(G) \neq \{1\}$.*

Proof a) The elements of M are partitioned into G-orbits, and the size of each orbit is a power of p by the Orbit-Stabiliser Theorem. Since fixed points are exactly the orbits of size 1, the claim follows.

b) This follows from a) applied to the conjugation action of G on itself. The elements of $Z(G)$ are exactly the fixed points under this action. Since the identity is a fixed point and $|G| \equiv 0 \bmod p$, there must be at least $p - 1$ further fixed points.

c) If $G/Z(G) = \langle gZ(G) \rangle$ for some $g \in G$, then every $h \in G$ can be written as $g^i z$ for some $i \in \mathbb{N}$ and $z \in Z(G)$. This implies that G is abelian, a contradiction.

d) We use induction on n. If G is abelian, which holds for $n \in \{0, 1\}$, then G is nilpotent. Thus, it remains to consider a non-abelian group G with $n \geqslant 2$. Part b) shows that $1 < Z(G) < G$ is non-trivial, and so $H = G/Z(G)$ is a finite p-group of order p^m for some m with $1 < m < n$ by c). Now the induction hypothesis implies that the upper central series of H terminates at H and has length at most $m - 1$. Since each $\zeta_i(H) = \zeta_{i+1}(G)/Z(G)$, it follows that the upper central series of G terminates at G and has length at most $m < n$. Thus, G is nilpotent of class less than n.

e) We use induction on n. If $n = 1$, then the claim is true. Now let $n > 1$. By b), the centre $Z(G)$ has a non-trivial element g. If $|g| = p^m$, then $h = g^{(p^{m-1})}$ has order p, hence $N = \langle h \rangle$ is a normal subgroup of size p and $H = G/N$ has order p^{n-1}. By the induction hypothesis, for every $i \in \{0, 1, \ldots, n-1\}$ there is a normal subgroup $U \leqslant H$ of order p^i. The full preimage of U under the projection $G \to H$ is a normal subgroup of size p^{i+1}. The claim follows.

f) Since G is non-trivial and nilpotent, there exists a smallest index i such that $Z = N \cap \zeta_i(G)$ is non-trivial. The definition of $\zeta_i(G)$ and the normality of N now imply that $[G, Z] \leqslant N \cap \zeta_{i-1}(G)$, and so $[G, Z] = \{1\}$ by the choice of i. Hence, Z is a non-trivial central subgroup. □

In preparation for the next result, we recall the following concepts. A subgroup M of G is *maximal* if M is a proper subgroup of G and $M \leqslant U \leqslant G$ implies that $U = M$ or $U = G$. If $N \trianglelefteq G$ is a normal subgroup, then the *Correspondence Theorem* shows that the subgroups of the quotient group G/N are exactly U/N where $U \leqslant G$ is a subgroup with $N \leqslant U$, and U/N is normal in G/N if and only if U is normal in G. The *normaliser* of a subgroup U in G is $N_G(U) = \{g \in G \mid u^g \in U \text{ for all } u \in U\}$; the *centraliser* is $C_G(U) = \{g \in G \mid ug = gu \text{ for all } u \in U\}$. Clearly, $C_G(U) \leqslant N_G(U) \leqslant G$ and $U \leqslant N_G(U)$; moreover, $U \cap C_G(U) = Z(U)$. A subgroup U of G

is *subnormal* in G if there exists a subgroup chain $U = U_0 \leqslant U_1 \leqslant \ldots \leqslant U_k = G$ such that each U_i is normal in U_{i+1}.

Proposition 1.5 *Let G be a non-trivial finite p-group.*

a) *Each proper subgroup U of G is a proper subgroup of its normaliser $N_G(U)$.*

b) *Each maximal subgroup M of G is normal in G with $[G : M] = p$.*

c) *Each subgroup U of G is subnormal in G.*

Proof a) We consider a proper subgroup $U < G$. Recall that $U \leqslant N_G(U)$. We show that this inclusion is proper. Let i be the smallest index such that $U\zeta_{i-1}(G) = U$ and $U\zeta_i(G) > U$; such an i exists since $G = \zeta_m(G)$ for some m and $UG = G > U$. Since $\zeta_i(G)/\zeta_{i-1}(G)$ is the centre of $G/\zeta_{i-1}(G)$, it follows from $U = U\zeta_{i-1}(G)$ that $\zeta_i(G) \leqslant N_G(U)$. This shows $U < N_G(U)$, as desired.

b) Let M be a maximal subgroup of G. By a) it follows that $M < N_G(M)$, and hence $N_G(M) = G$. Thus, M is normal in G, and G/M is a quotient group of G. As M is maximal, G/M is non-trivial, and the Correspondence Theorem implies that the only subgroups are G/M and M/M. Proposition 1.4e) now shows that G/M has order p, so $[G : M] = p$ as claimed.

c) This follows directly from a). □

The *Frattini subgroup* $\Phi(G)$ of a group G is the intersection of all maximal subgroups of G with the stipulation that $\Phi(G) = G$ if G has no maximal subgroups. If $X \subseteq G$ is a subset, then $\langle X \rangle$ denotes the subgroup of G generated by X. Relying on the Axiom of Choice, the next result shows that the Frattini subgroup is the set of *non-generators* in a group, that is, the set of elements that are redundant in every generating set of G.

Proposition 1.6 *Let G be a group. Then $g \in \Phi(G)$ if and only if $G = \langle X, g \rangle$ implies $G = \langle X \rangle$ for all $X \subseteq G$.*

Proof First consider $g \in \Phi(G)$, and let $X \subseteq G$ with $G = \langle X, g \rangle$. Suppose, for a contradiction, that $G \neq \langle X \rangle$. Let $\mathcal{M}$ be the set of all subgroups $M < G$ that contain X but not g. This set is partially ordered by inclusion, and the union of any chain $M_1 \subseteq M_2 \subseteq \ldots$ of subgroups in $\mathcal{M}$ lies in $\mathcal{M}$. Zorn's Lemma (equivalent to the Axiom of Choice) now asserts that there exists a maximal element of $\mathcal{M}$, that is, $M < G$ that is maximal with respect to the property that $X \subseteq M$ and $g \notin M$. This subgroup is a maximal subgroup of G: If $M < H \leqslant G$, then the choice of M implies $g \in H$ and $X \subseteq H$, hence $G = \langle M, g \rangle = H$. As M is maximal, $g \in \Phi(G) \leqslant M$. This is a contradiction to $g \notin M$. Thus, $G = \langle X \rangle$, and so g is a non-generator.

Now suppose that $g \notin \Phi(G)$. Then there exists a maximal subgroup M of G with $g \notin M$. Now $\langle M, g \rangle = G$ and $M < G$ shows that g is not a non-generator. □

For $n \in \mathbb{N}$ we write $G^{[n]} = \langle g^n \mid g \in G \rangle$ for the subgroup generated by the n-th powers of elements. The n-fold direct product of copies of G is denoted G^n, and C_n is the cyclic group of order n. Recall that a group is elementary abelian if it is isomorphic to C_p^n for some prime p and integer n.

Proposition 1.7 *The Frattini subgroup of a finite p-group G is $\Phi(G) = G'G^{[p]}$. Equivalently, $\Phi(G)$ is the smallest normal subgroup of G with elementary abelian factor group.*

Proof If M is a maximal subgroup of G, then $G/M \cong C_p$ is cyclic of order p by Proposition 1.5, and therefore $G'G^{[p]} \leqslant M$. This proves $G'G^{[p]} \leqslant \Phi(G)$. For the other inclusion, observe that $G/G'G^{[p]}$ is elementary abelian by construction, and hence its Frattini subgroup is trivial. This implies that the maximal subgroups of G intersect in a subgroup of $G'G^{[p]}$, that is, $\Phi(G) \leqslant G'G^{[p]}$. □

There are different notions of rank for a group G. In this work, the *rank* $\mathrm{rk}(G)$ of a group G denotes the smallest possible size of a generating set for G. For example, the elementary abelian group C_p^n has rank n. The previous results imply that the rank of a finite p-group can be read off from its Frattini quotient.

Corollary 1.8 *If G is a finite p-group, then $G/\Phi(G)$ is elementary abelian of rank* $\mathrm{rk}(G)$.

Proof Proposition 1.7 shows that the Frattini quotient is elementary abelian, say $G/\Phi(G) \cong C_p^r$. This implies that $\mathrm{rk}(G/\Phi(G)) = r$ and it remains to show that $\mathrm{rk}(G) = r$. Let $\{g_1\Phi(G), \ldots, g_r\Phi(G)\}$ be a generating set of $G/\Phi(G)$. Then $G = \langle g_1, \ldots, g_r, \Phi(G)\rangle$, and an iteration of Proposition 1.6 shows that $G = \langle g_1, \ldots, g_r\rangle$, hence $\mathrm{rk}(G) \leqslant r$. On the other hand, each generating set of G induces a generating set of $G/\Phi(G)$, and therefore $\mathrm{rk}(G) \geqslant \mathrm{rk}(G/\Phi(G)) = r$. □

To conclude this section, we consider the automorphism group $\mathrm{Aut}(G)$ of a finite p-group G of rank r. Note that $\Phi(G)$ is characteristic in G and thus every automorphism α of G induces an automorphism $\alpha_{G/\Phi(G)}$ of $G/\Phi(G)$ defined by $g\Phi(G) \mapsto (g^\alpha)\Phi(G)$. As $G/\Phi(G) \cong C_p^r$, it follows that $G/\Phi(G)$ can be considered as an r-dimensional vector space over the field with p elements, and thus $\mathrm{Aut}(G/\Phi(G) \cong \mathrm{GL}_r(p)$, the group of invertible $r \times r$ matrices over the field with p elements. This is exploited further in the following proposition.

Proposition 1.9 *Let G be a p-group of order p^n and rank r. The kernel of the map $\sigma\colon \mathrm{Aut}(G) \to \mathrm{Aut}(G/\Phi(G))$, $\alpha \mapsto \alpha_{G/\Phi(G)}$, is a p-group of order dividing $p^{(n-r)r}$, and the image of σ is isomorphic to a subgroup of* $\mathrm{GL}_r(p)$.

Proof By construction, the image of the homomorphism σ is a subgroup of $\mathrm{Aut}(G/\Phi(G)) \cong \mathrm{GL}_r(p)$; it remains to investigate the kernel $K = \ker\sigma$. Let $(g_1, \ldots, g_r)$ be an ordered generating set of G, and let $\mathcal{M}$ be the set of all ordered r-tuples of the form $(g_1 f_1, \ldots, g_r f_r)$ with each $f_i \in \Phi(G)$. By Proposition 1.6, each element of $\mathcal{M}$ is an ordered generating set of G. Since every automorphism α of G is determined by the images of a generating set, it follows that $\alpha \in K$ if and only if $(h_1^\alpha, \ldots, h_r^\alpha) \in \mathcal{M}$ for every $(h_1, \ldots, h_r) \in \mathcal{M}$. This defines a group action of K on $\mathcal{M}$ with the property that K acts regularly on each K-orbit in $\mathcal{M}$. It follows that every orbit has size $|K|$, and therefore $|K|$ divides $|\mathcal{M}| = p^{(n-r)r}$. □

1.2 The definition of coclass

Proposition 1.4 shows that every finite p-group is nilpotent and hence has a class. This leads to the following fundamental definition.

Definition 1.10 (Leedham-Green & Newman [60]). The *coclass* of a finite p-group G of order p^n is defined as $\mathrm{cc}(G) = n - \mathrm{cl}(G)$.

The groups of order 1 and p have coclass 0; for larger orders the following holds by Proposition 1.4d).

Corollary 1.11 *If G has order p^n with $n \geqslant 2$, then* $\mathrm{cc}(G) \in \{1, 2, \dots, n-1\}$.

The p-groups of coclass 1 have *maximal class*. Following Wiman [82], the first major study of the p-groups of coclass 1 was carried out by Blackburn [1], whose work includes a classification of the 2-groups of coclass 1 up to isomorphism. There are two such groups of order 2^2, namely the cyclic group of order 4 and the elementary abelian group of order 4, there are two such groups of order 2^3, namely the dihedral group of order 8 and the quaternion group of order 8, and there are three such groups of order 2^n for each $n \geqslant 4$, namely the dihedral group of order 2^n, the quaternion group of order 2^n and the semidihedral group of order 2^n. For $n \geqslant 4$, these groups (each of order 2^n) can be described by the following group presentations:

$$\begin{aligned} D_{2^n} &= \langle a, b \mid a^2 = 1, \quad b^{2^{n-1}} = 1,\ b^a = b^{-1}\rangle, \\ SD_{2^n} &= \langle a, b \mid a^2 = 1, \quad b^{2^{n-1}} = 1,\ b^a = b^{2^{n-2}-1}\rangle, \\ Q_{2^n} &= \langle a, b \mid a^2 = b^{2^{n-2}},\ b^{2^{n-1}} = 1,\ b^a = b^{-1}\rangle. \end{aligned} \tag{1.1}$$

Blackburn also classified the finite 3-groups of coclass 1. If $n = 2m > 5$ is even, then there are seven (isomorphism types of) groups of order 3^n and coclass 1. These seven isomorphism types correspond to the parameters $(u, v, w) \in \{(0,0,0), (0,0,1), (0,0,2), (0,1,0), (0,2,0), (1,1,0), (1,1,1)\}$ and can be defined as the groups with abstract generators a, b, c, d and relations

$$a^3 = d^u,\ b^{3^{m-1}} = d,\ c^{3^{m-1}} = d,\ d^3 = 1,$$

$$b^a = c,\ c^a = b^{3^{m-1}-1} c^{3^{m-1}-1} d^v,\ c^b = cd^w,\ d \text{ central}.$$

If $n = 2m + 1 \geqslant 5$ is odd, then there are six groups of order 3^n with coclass 1. Using $(u, v, w) \in \{(0,0,0), (0,0,1), (0,0,2), (0,2,0), (1,0,1), (1,2,0)\}$, these groups can be defined as the groups with abstract generators a, b, c, d and relations

$$a^3 = d^u,\ b^{3^{m-1}} = c^{3^{m-1}} d,\ c^{3^m} = 1,\ d^3 = 1,$$

$$b^a = c,\ c^a = b^{3^{m-1}-1} c^{2\cdot 3^{m-1}-1} d^v,\ c^b = cd^w,\ d \text{ central}.$$

The classification of the finite 5-groups of coclass 1 is listed as Problem 3 in Shalev's survey paper [77] on finite p-groups. Their classification and that of the p-groups of coclass 1 for primes $p \geqslant 7$ remains open, although significant partial results exist.

The next result is a first indication that the coclass of a group has a significant bearing on the structure of a group.

Lemma 1.12 *If G is a finite p-group with coclass r, then $|G/\gamma_2(G)| \leqslant p^{r+1}$ and $|Z(G)| \leqslant p^{r+1}$. In particular, G has rank at most $r+1$.*

Proof If G has class c and $|G/\gamma_2(G)| = p^m$, then $|G| \geqslant p^{m+c-1}$ and $r \geqslant m-1$. Similarly, if $Z(G) = \zeta_1(G)$ has order p^m, then $|G| \geqslant p^{m+c-1}$, so $r \geqslant m-1$. □

In general, coclass does not behave well when forming subgroups or quotient groups, as illustrated by the cyclic group of order p^k: this group has subgroups and quotients of coclass $0, 1, \ldots, k-1$, respectively. If $n \geqslant 2$, then D_{2^n} has coclass 1 and there are cyclic subgroups with coclass $0, 1, \ldots, n-2$, respectively.

Proposition 1.13 *Let G and H be two finite p-groups of coclass r and let $\varphi\colon G \to H$ be an epimorphism.*

a) *The kernel of φ is $\gamma_{\mathrm{cl}(H)+1}(G)$.*

b) *The quotient $G/\gamma_j(G)$ has coclass r for all $j \geqslant \mathrm{cl}(H)+1$.*

c) *If $\psi\colon G \to H$ is another epimorphism, then $\psi = \varphi\beta$ for a unique $\beta \in \mathrm{Aut}(H)$.*

Proof a) Let $|G| = p^n$ and $|H| = p^m$. As $\mathrm{cc}(G) = \mathrm{cc}(H)$, it follows that $n - \mathrm{cl}(G) = m - \mathrm{cl}(H)$ and thus $n - m = \mathrm{cl}(G) - \mathrm{cl}(H)$. Since $(\gamma_i(G))^\varphi = \gamma_i(H)$ for all i,

$$\ker\varphi \geqslant \gamma_{\mathrm{cl}(H)+1}(G) > \ldots > \gamma_{\mathrm{cl}(H)+n-m+1}(G) = \gamma_{\mathrm{cl}(G)+1} = \{1\},$$

and now $\ker\varphi = \gamma_{\mathrm{cl}(H)+1}(G)$ follows from $|\ker\varphi| = p^{n-m}$.

b) The proof of a) shows that $\gamma_{\mathrm{cl}(H)+1}(G) > \ldots > \gamma_{\mathrm{cl}(G)+1}(G) = \{1\}$ is a properly descending series with $\mathrm{cl}(G) - \mathrm{cl}(H)$ steps, and therefore each quotient $\gamma_i(G)/\gamma_{i+1}(G)$ of this series has order p. Thus, $G/\gamma_j(G)$ has coclass $\mathrm{cc}(G)$ for each $j \in \{\mathrm{cl}(H)+1, \ldots, \mathrm{cl}(G)+1\}$.

c) Part a) shows that φ and ψ have the same kernel $N = \gamma_{\mathrm{cl}(H)+1}(G)$. By the first Isomorphism Theorem, $\varphi'\colon G/N \to H, gN \mapsto g^\varphi$, and $\psi'\colon G/N \to H, gN \mapsto g^\psi$, are isomorphisms, and therefore $\beta\colon H \to H$, $h \mapsto h^{(\varphi')^{-1}\psi'}$, is an automorphism. By construction, $\psi = \varphi\beta$, and β is unique. □

1.3 Introduction to pro-p-groups

Infinite pro-p-groups play a crucial role in coclass theory. In the literature, pro-p-groups and, more generally, pro-finite groups are often approached via topology. Here we use an algebraic way towards their analysis and quickly focus on the infinite pro-p-groups of coclass r. In doing so, we supply complete proofs of certain key results that in the existing literature often appear only as exercises.

For an in-depth discussion of pro-p-groups we refer to Dixon et al. [19] or Leedham-Green & McKay [57].

Let $(\Lambda, \geqslant)$ be a partially ordered set such that for every $i, j \in \Lambda$ there exists some $k \in \Lambda$ with $k \geqslant i$ and $k \geqslant j$. An *inverse system* of finite groups over Λ is a family of finite groups $\{G_i \mid i \in \Lambda\}$, together with group epimorphisms $\pi_{i,j}\colon G_i \to G_j$ whenever $i \geqslant j$, such that $\pi_{i,i} = \mathrm{id}_{G_i}$ for all $i \in \Lambda$ and $\pi_{i,k} = \pi_{i,j}\pi_{j,k}$ whenever $i, j, k \in \Lambda$ with $i \geqslant j \geqslant k$. The *inverse limit* of this system is the subgroup

$$G = \varprojlim (G_i)_{i \in \Lambda}$$

of the Cartesian product $\prod_{i \in \Lambda} G_i$ that consists of all elements $(g_i)_{i \in \Lambda}$ satisfying $g_i^{\pi_{i,j}} = g_j$ whenever $i, j \in \Lambda$ with $i \geqslant j$.

A special case is a *linear inverse system* where Λ is countable and ordered like $\mathbb{N}$. In this case an inverse system is based on a sequence $G_1, G_2, \ldots$ of finite groups together with epimorphisms $\pi_i\colon G_{i+1} \to G_i$ for each $i \geqslant 1$. Epimorphisms $G_i \to G_j$ for $i > j+1$ are implicitly defined as the composition $\pi_{i-1}\pi_{i-2}\ldots\pi_j$. The corresponding inverse limit is

$$\varprojlim G_i = \{(g_1, g_2, \ldots) \mid g_j \in G_j \text{ and } g_{j+1}^{\pi_j} = g_j \text{ for all } j \geqslant 1\}.$$

The inverse limit of finite groups is a *pro-finite group*; the following is a standard example.

Example 1.14 Let p be a prime, and for each integer $i \geqslant 1$ let $G_i = \mathbb{Z}/p^i\mathbb{Z}$ be the cyclic group of order p^i written additively. Denote by $\pi_i\colon G_{i+1} \to G_i$ the natural epimorphism that maps $1 + p^{i+1}\mathbb{Z}$ to $1 + p^i\mathbb{Z}$. The inverse limit of this system is the additive group of the ring of *p-adic integers*, that is,

$$\mathbb{Z}_p = \varprojlim (\mathbb{Z}/p^i\mathbb{Z}).$$

The elements of $\mathbb{Z}_p$ are sequences $z = (z_i)_{i \geqslant 1}$ with each $z_j \in \{0, 1, \ldots, p^j - 1\}$ such that $z_j \equiv z_k \bmod p^j$ for each $k \geqslant j$.

We now turn to those inverse limits that are of interest in this book.

Definition 1.15 A *pro-p-group* is the inverse limit of an inverse system of finite p-groups. A pro-p-group has coclass r if it is the inverse limit of an inverse system of finite p-groups of coclass r.

Example 1.16 We consider the dihedral groups D_{2^i} as defined in (1.1). We define an epimorphism $\pi_i\colon D_{2^{i+1}} \to D_{2^i}$ via $a^{\pi_i} = a$ and $b^{\pi_i} = b$. The inverse limit D of this linear inverse system is a pro-2-group of coclass 1. Since each D_{2^i} contains a cyclic group of order 2^{i-1} as normal subgroup of index 2, the group D contains a normal subgroup of index 2, isomorphic to the additive group of the 2-adic integers $\mathbb{Z}_2$. The subgroup generated by a in D_{2^i} has order 2 and acts by inversion on b, which translates to multiplication by -1 in the 2-adic integers $\mathbb{Z}_2$. Induction on j can be used to prove that $\gamma_j(D) = 2^{j-1}\mathbb{Z}_2$ for all $j \geqslant 2$, and so $D/\gamma_j(D) \cong D_{2^j}$. We note that D is uncountable and hence not isomorphic to the infinite dihedral group $D_\infty = \langle a, b \mid a^2, b^a = b^{-1}\rangle$.

An important consequence of Proposition 1.13 is that every pro-p-group of coclass r can be defined via a linear inverse system. As an illustration of working with these definitions, we include full details for the proof.

Proposition 1.17 *Let G be the inverse limit of the inverse system $\Gamma = \{G_i \mid i \in \Lambda\}$ of finite p-groups of coclass r, with epimorphisms $\pi_{i,j}$ whenever $i \geqslant j$ in Λ.*

a) *If $i, j \in \Lambda$ with $|G_i| \leqslant |G_j|$, then there is an epimorphism $\tau\colon G_j \to G_i$ such that $\pi_{k,i} = \pi_{k,j}\tau$ for every $k \in \Lambda$ with $k \geqslant i, j$. If $|G_i| = |G_j|$, then $G_i \cong G_j$.*

b) *The group G is isomorphic to the inverse limit of a linear inverse system.*

Proof a) Let $k \in \Lambda$ with $k \geqslant i, j$, which exists by the definition of an inverse system. Proposition 1.13 shows that $\ker \pi_{k,j}$ is contained in $\ker \pi_{k,i}$, thus τ can be defined as follows: If $g \in G_j$, then choose a preimage $h \in G_k$ under $\pi_{k,j}$ and define $g^\tau = h^{\pi_{k,i}}$. It remains to show that τ is independent of the choice of k. Suppose $k' \in \Lambda$ with $k' \geqslant i, j$, and let $\tau'\colon G_j \to G_i$ be the epimorphism that is constructed with respect to k'. There exists $\ell \in \Lambda$ with $\ell \geqslant k, k'$, which leads to the diagram of epimorphisms displayed in Figure 1.1 (left): the diagram of solid arrows is commutative, and we have $\pi_{k,i} = \pi_{k,j}\tau$ and $\pi_{k',i} = \pi_{k',j}\tau'$. If $x \in G_j$ and we choose $y \in G_\ell$ as a preimage of x under $\pi_{\ell,j}$, then $x^\tau = y^{\pi_{\ell,i}} = x^{\tau'}$. This holds for every $x \in G_j$, so $\tau = \tau'$.

b) The claim is trivially true if there is a bound on the orders of the groups in Γ, so assume in the following that there is no such bound. We construct a linear inverse system $\Gamma' = \{G_{i_j} \mid j = 1, 2, \ldots\}$ contained in Γ. Initialise $\Gamma' = \{G_{i_1}\}$ where G_{i_1} is any group in Γ. Suppose now Γ' is extended to $\Gamma' = \{G_{i_1}, \ldots, G_{i_n}\}$ with $i_n \geqslant \ldots \geqslant i_1$ in Λ. By our assumption that the orders of the groups in Γ are unbounded, we can choose $G_\ell \in \Gamma$ with $|G_\ell| > |G_{i_n}|$. If $\ell \geqslant i_n$ in Λ, then define $G_{i_{n+1}} = G_\ell$. Otherwise, there exists $k \in \Lambda$ with $k \geqslant \ell$ and $k \geqslant i_n$, and we define $G_{i_{n+1}} = G_k$. An iteration of this process yields a linear inverse system Γ' whose groups and epimorphisms come from the original inverse system Γ. Write $\Lambda' = \{i_1, i_2, \ldots\}$, so that $\Gamma' = \{G_j \mid j \in \Lambda'\}$, and let H be the inverse limit of Γ'. Since $\Lambda' \subseteq \Lambda$, we can define the homomorphism

$$\pi\colon G \to H, \quad (g_i)_{i\in\Lambda} \mapsto (g_i)_{i\in\Lambda'}.$$

We show in the remainder of the proof that π is an isomorphism.

First, consider $g = (g_i)_{i\in\Lambda}$ in the kernel of π. Let $i \in \Lambda$ and choose $j \in \Lambda'$ such that $|G_i| \leqslant |G_j|$. Let $k \in \Lambda$ with $k \geqslant i$ and $k \geqslant j$, and let $\tau\colon G_j \to G_i$ be the epimorphism of part a) with $\pi_{k,i} = \pi_{k,j}\tau$. By assumption, $g \in \ker \pi$ and $j \in \Lambda'$, so $g_j = 1$. Thus, $g_i = g_k^{\pi_{k,i}} = g_k^{\pi_{k,j}\tau} = g_j^\tau = 1$, and so $g_i = 1$ for every $i \in \Lambda$. This shows that π is injective.

We note that establishing surjectivity is not trivial, despite the fact that π is defined via a projection; we refer to Ribes and Zalesskii [70, p. 6] for a further discussion.

To prove surjectivity of π we first construct homomorphisms $\sigma_i\colon H \to G_i$ for $i \in \Lambda$. For a given i let $j \in \mathbb{N}$ be minimal with $|G_{i_j}| \geqslant |G_i|$. Denote by $\tau_i\colon G_{i_j} \to G_i$ the epimorphism of part a), and let $\rho_{i_j}\colon H \to G_{i_j}$ be the projection that maps $(g_{i_j})_{j\in\mathbb{N}}$ to g_{i_j}. We use these maps to define the homomorphism

$$\sigma_i\colon H \longrightarrow G_i, \quad h \mapsto h^{\rho_{i_j}\tau_i}.$$

If $i \in \Lambda'$, then we take $i_j = i$, so that $\sigma_i = \rho_i$.
For a given $h = (h_{i_j})_{j\in\mathbb{N}} \in H$ define $h' = (h^{\sigma_i})_{i\in\Lambda}$. We claim that h' is a preimage of h under π, which will show surjectivity of π. This follows by construction if we have shown that h' actually lies in G. To prove the latter, we need to consider $i' \geqslant i$ in Λ and show that $h^{\sigma_{i'}\pi_{i',i}} = h^{\sigma_i}$. Suppose $\sigma_{i'}$ and σ_i are constructed as $\sigma_{i'} = \rho_{i_{j'}}\tau_{i'}$ and $\sigma_i = \rho_{i_j}\tau_i$. Note that $i_{j'} \geqslant i_j$, so h_{i_j} is the image of $h_{i_{j'}}$ under $\pi_{i_{j'},i_j}$. Now choose $k \in \Lambda$ with $k \geqslant i, i_j, i', i_{j'}$, so that $\pi_{k,i} = \pi_{k,i_j}\tau_i$ and $\pi_{k,i'} = \pi_{k,i_{j'}}\tau_{i'}$ by a). With this set-up, showing that $h^{\sigma_{i'}\pi_{i',i}} = h^{\sigma_i}$ is equivalent to showing that

$$h_{i_{j'}}^{\tau_{i'}\pi_{i',i}} = h_{i_{j'}}^{\pi_{i_{j'},i_j}\tau_i}.$$

That the latter holds follows from the diagram in Figure 1.1 (right) where the subdiagram of solid arrows is commutative, together with $\pi_{k,i} = \pi_{k,i_j}\tau_i$ and $\pi_{k,i'} = \pi_{k,i_{j'}}\tau_{i'}$. In conclusion, $h' \in G$, and so π is an isomorphism. □

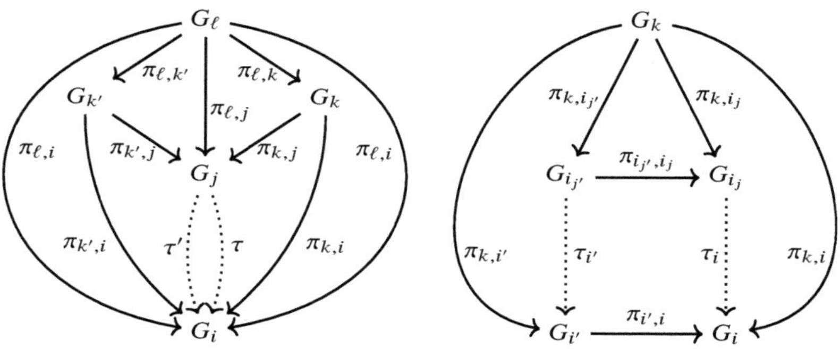

Fig. 1.1 Two commutative diagrams illustrating the proof of Proposition 1.17

Motivated by Proposition 1.17, we exhibit some elementary facts about pro-p-groups of coclass r defined by linear inverse limits.

Proposition 1.18 *Let G be the inverse limit of the inverse system $G_1, G_2, \ldots$ of finite p-groups of coclass r, with epimorphisms $\pi_i\colon G_{i+1} \longrightarrow G_i$.*

a) *The projection $G \longrightarrow G_i$, $(g_1, g_2, \ldots) \mapsto g_i$, is surjective for every i.*
b) *If there is no bound on the orders of the groups $G_1, G_2, \ldots$, then G is infinite; otherwise, there exists an index j such that $G \cong G_j$ is finite.*
c) *Let $j_1 < j_2 < \ldots$ be an infinite sequence of integers. Let H be the inverse limit of the inverse system $G_{j_1}, G_{j_2}, \ldots$ with epimorphisms $\pi'_{j_i} = \pi_{j_{i+1}-1}\pi_{j_{i+1}-2}\cdots\pi_{j_i}$. The projection $(g_1, g_2, \ldots) \mapsto (g_{j_1}, g_{j_2}, \ldots)$ is an isomorphism $G \longrightarrow H$.*

Proof a) Let $g_i \in G_i$. For $j < i$ set $g_j = g_i^{\pi_{i-1}\cdots\pi_j}$, and for $k > i$ iteratively choose g_k as a preimage of g_{k-1}. Then $g = (g_1, g_2, \ldots)$ is a preimage of g_i in G.

b) If the orders are unbounded, then part a) proves that G is infinite. Otherwise, choose j such that $|G_j| \geqslant |G_m|$ for all m. Since all the π_i are epimorphisms, the choice of j implies that π_i is an isomorphism whenever $i > j$. If $g = (g_1, g_2, \ldots) \in G$, then each g_i with $i \leqslant j$ is uniquely determined as $g_j^{\pi_{j-1}\ldots\pi_i}$. Conversely, each g_i with $i > j$ is uniquely determined as the preimage under the isomorphism $\pi_{i-1}\ldots\pi_j$. Thus, $G \to G_j$, $(g_1, g_2, \ldots) \mapsto g_j$ is injective. It is surjective by a).

c) This is a special case of the discussion in the proof of Proposition 1.17. □

Another interesting feature of linear inverse limits of finite p-groups of coclass r and another consequence of Proposition 1.13 is that the isomorphism type is independent of the chosen epimorphisms.

Proposition 1.19 *Consider two linear inverse systems of finite p-groups $G_1, G_2, \ldots$ of coclass r with epimorphisms $\pi_i\colon G_{i+1} \to G_i$ and $\pi_i'\colon G_{i+1} \to G_i$, respectively. The inverse limit with respect to $\pi_1, \pi_2, \ldots$ is isomorphic to the inverse limit with respect to $\pi_1', \pi_2', \ldots$.*

Proof For $i \geqslant 1$ abbreviate $A_i = \mathrm{Aut}(G_i)$. For $i \geqslant j$ let $\pi_{i,j} = \pi_{i-1}\pi_{i-2}\ldots\pi_j$ and $\pi_{i,j}' = \pi_{i-1}'\pi_{i-2}'\ldots\pi_j'$ be the induced epimorphisms $G_i \to G_j$. Proposition 1.13 shows that for each $i \geqslant 1$ we can decompose $\pi_i' = \pi_i\beta_i$ for some $\beta_i \in A_i$. It also shows that for every $m \geqslant i \geqslant 1$ there is an automorphism $\beta_{m,i} \in A_i$ with $\beta_m\pi_{m,i}' = \pi_{m,i}\beta_{m,i}$. Since A_1 is finite, there exists $\alpha_1 \in A_1$ such that $\mathcal{I}_1 = \{m \in \mathbb{N} \mid \beta_{m,1} = \alpha_1\}$ is an infinite set. By the same argument, there exists some automorphism $\alpha_2 \in A_2$ such that $\mathcal{I}_2 = \{m \in \mathcal{I}_1 \mid \beta_{m,2} = \alpha_2\}$ is infinite. An iteration of this process yields a sequence of automorphisms $\alpha_1, \alpha_2, \ldots,$ such that for every $i \geqslant 1$ and every $m \in \mathcal{I}_i$ the subdiagram of solid lines in Figure 1.2 is commutative. This implies that also $\pi_i\alpha_i = \alpha_{i+1}\pi_i'$ for all $i \geqslant 1$, which shows that the map defined by $(g_1, g_2, \ldots) \mapsto (g_1^{\alpha_1}, g_2^{\alpha_2}, \ldots)$ yields an isomorphism between the two inverse limits in the proposition. □

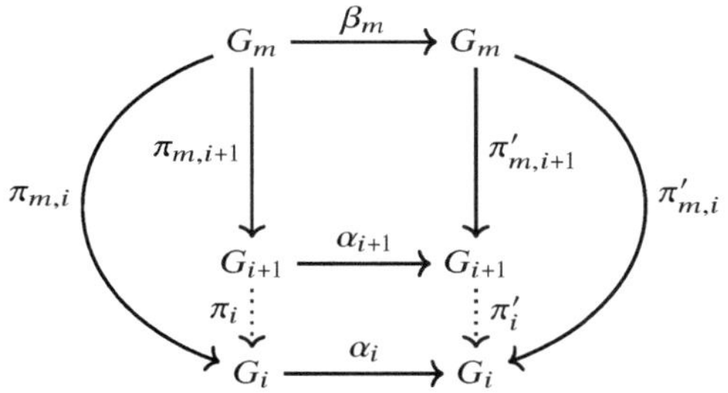

Fig. 1.2 A commutative diagram illustrating the proof of Proposition 1.19

Lastly, we mention that a pro-p-group G of coclass r is isomorphic to the inverse limit of its nilpotent quotients $G/\gamma_2(G), G/\gamma_3(G), \ldots$. Proposition 1.19 shows that the isomorphism type of this inverse limit is independent of the chosen projections $G/\gamma_{i+1}(G) \to G/\gamma_i(G)$.

Proposition 1.20 *Let G be a pro-p-group of coclass r. There exists an integer $k \geqslant 2$ such that $G/\gamma_j(G)$ is a finite p-group of coclass r for every $j \geqslant k$. Moreover, G is isomorphic to every inverse limit of $G/\gamma_2(G)$, $G/\gamma_3(G)$, ...*

Proof By Proposition 1.17, we can assume that G is the inverse limit of a linear inverse system $G_1, G_2, \dots$ of finite groups of coclass r, with epimorphisms $\pi_i\colon G_{i+1} \to G_i$. Let $k = \mathrm{cl}(G_1) + 1$ and $j \geqslant 2$. By Proposition 1.13, the induced projection $G_j \to G_1$ has kernel $\gamma_k(G_j)$ and $G_j/\gamma_\ell(G_j)$ has coclass r for $\ell \geqslant k$. Now let $j \geqslant k$, and denote by N_j the inverse limit of the groups $\gamma_j(G_1), \gamma_j(G_2), \dots$ where the epimorphisms are the ones induced by the π_i. By Proposition 1.18, each projection $G \to G_\ell$ is surjective and maps $\gamma_j(G)$ to $\gamma_j(G_\ell)$. It follows that $N_j = \gamma_j(G)$, and therefore $G/\gamma_j(G) \cong G_\ell/\gamma_j(G_\ell)$ for all G_ℓ with $\mathrm{cl}(G_\ell) + 1 \geqslant j \geqslant k$. Since $G_\ell/\gamma_j(G_\ell)$ has coclass r, the first claim follows. It also follows that $G_\ell \cong G/\gamma_{\mathrm{cl}(G_\ell)+1}(G)$ for every $\ell \geqslant 1$, so G is isomorphic to the inverse limit of all $G/\gamma_{\mathrm{cl}(G_\ell)+1}(G)$ with $\ell \geqslant 1$. By Proposition 1.18, the group G is then also isomorphic to the inverse limit of all $G/\gamma_i(G)$, as claimed. Proposition 1.19 explains that a different choice of epimorphisms yields an isomorphic group. □

We conclude this section with a brief comment on the topological approach towards pro-p-groups of finite coclass. Some of these results will be used in a few proofs discussed in Chapter 4.

Remark 1.21 Let G be the inverse limit of an inverse system $\{G_i \mid i \in \Lambda\}$ of finite groups, and furnish each G_i with the discrete topology. The Cartesian product $\prod_{i\in\Lambda} G_i$ then carries the product topology and its subgroup G the induced topology. Thus, G is a *topological group*, that is, a group that is also a topological space such that the map $G \times G \to G\colon (g, h) \mapsto gh^{-1}$ is continuous. Topology plays a significant role in many investigations of pro-finite groups. We refer to Dixon et al. [19] and Ribes & Zalesskii [70] for many details; in particular, see [19, pp. 16, 21-27, 245, 267]. Here we include a brief summary of some relevant results. For a subgroup U of a topological group G denote by $\overline{U}$ the (topological) closure of U. Further, note that G is finitely generated as a topological group if there exists a finite set $X \subseteq G$ with $G = \overline{\langle X \rangle}$.

a) According to the standard topological definition, a pro-p-group G has coclass r if almost all quotients $G/\overline{\gamma_i(G)}$ are finite p-groups of coclass r. Every such group G is finitely generated as a topological group. The algebraic Definition 1.15 always yields a pro-p-group that has coclass r in this topological sense.

b) A closed subgroup of a pro-p-group is a pro-p-group; the quotient of a pro-p-group by a closed normal pro-p-subgroup is a pro-p-group. A pro-p-group G that is finitely generated as a topological group satisfies the following: the open subgroups of G are exactly the subgroups of finite index; every open subgroup is finitely generated as a topological group; every abstract group homomorphism from G to a pro-p-group is continuous; the topology of G is determined by the group structure of G; the abstract groups $\Phi(G)$ and $\gamma_i(G)$ are closed.

c) If G is a pro-p-group such that almost all $G/\overline{\gamma_i(G)}$ are finite p-groups of coclass r, then each of $\Phi(G)$ and $\gamma_i(G)$ is both open and closed. Together with a,b), it follows that the algebraic definition of coclass (Definition 1.15) is equivalent to the topological one.

d) If U is a subgroup of a pro-p-group G of coclass r, then $\overline{U} = \bigcap_{j\in\mathbb{N}} U\gamma_j(G)$.

1.4 The *Coclass Conjectures*

Leedham-Green & Newman [60] proposed five conjectures, known as Coclass Conjectures A–E, on the structure of the p-groups of a fixed coclass. These conjectures can be seen as the origin of coclass theory, and an important milestone in this project was the complete proof of all five conjectures. We state the Coclass Conjectures as theorems and include comments and references to their proofs below. For a recent discussion of the history of conjectures we refer to [16].

Coclass Theorem A. *There is a function $f(p,r)$ such that every finite p-group of coclass r has a normal subgroup of class at most 2 and index at most $f(p,r)$. If $p = 2$, then the normal subgroup can be taken to be abelian.*

This is the strongest of the five conjectures. Leedham-Green [52] and Shalev [76] give independent proofs. The proof by Shalev yields explicit bounds for $f(p,r)$: it shows that

$$f(p,r) \leqslant \begin{cases} p^{2(p-1)p^{r-1}+r-3} & (p \text{ odd}), \\ 2^{3.2^{r+1}+r-2} & (p = 2), \end{cases}$$

we refer to [57, Theorems 6.4.5 and 6.4.3] for a proof. The proof by Leedham-Green is not fully self-contained, but introduces various highly interesting ideas along the way; some of these play major roles in this book. Both proofs are described in [57].

Coclass Theorem B. *There is a function $g(p,r)$ such that every finite p-group of coclass r has derived length at most $g(p,r)$.*

This follows from Theorem A: Let G be a finite p-group of coclass r and let $N \trianglelefteq G$ be a normal subgroup of class at most 2 and index $f(p,r)$. Then $H = G/N$ is a finite p-group of order at most $f(p,r)$ and hence has derived length at most $f(p,r) - 1$. Thus the derived length of G is at most $f(p,r) + 1$. These bounds can be made explicit by using the bounds for $f(p,r)$ as above. First, consider p odd. As H has coclass at most r, it has class at most $2(p-1)p^{r-1}-3$. If $i = \lceil \log_2(2(p-1)p^{r-1}-2) \rceil$, then $\gamma_{2^i}(H) = 1$, hence $G^{(i)} \leqslant \gamma_{2^i}(G) \leqslant N$. Since N has class at most 2, we have $G^{(i+2)} = \{1\}$. Thus, G has derived length at most $4 + \log_2((p-1)p^{r-1} - 1)$. The argument for $p = 2$ is analogous and yields an upper bound $r + 4$.

Coclass Theorem C. *Every pro-p-group of coclass r is soluble.*

Proofs were given by Donkin [20] for primes $p > 3$, see also Leedham-Green [51], and by Shalev & Zelmanov [78] for all primes. Here we briefly show how Theorem B implies Theorem C. Let G be an infinite pro-p-group of coclass r such that $G/\gamma_j(G)$ is a finite p-group of coclass r for all $j \geqslant j_0$. By Theorem B, the derived length of

$G/\gamma_j(G)$ is at most $g(p,r)$, and this bound depends on p and r, but not on j. This shows that with finitely many exceptions, the terms of the derived series of G lie in $\bigcap_{j\geqslant 2}\gamma_j(G)$. It now remains to note that $\bigcap \gamma_j(G) = \{1\}$. This follows as G can be defined as the inverse limit of a linear inverse system of finite p-groups $G_1, G_2, \ldots$ of coclass r, and for every $m \geqslant 0$ there exists an integer $j(m) \geqslant 0$ such that if $(g_1, g_2, \ldots) \in \gamma_{j(m)}(G)$, then $g_1, g_2, \ldots, g_m$ are all trivial.

Coclass Theorem D. *There are only finitely many isomorphism types of infinite pro-p-groups of coclass r.*

This theorem follows from Theorem C and some structure results for soluble pro-p-groups of finite coclass. We omit details here and refer the interested reader to [19, Sections 10.1 and 10.2]. The last theorem is a direct consequence of Theorem D.

Coclass Theorem E. *There are only finitely many isomorphism types of infinite soluble pro-p-groups of coclass r.*

1.5 The definition of the coclass graph $\mathcal{G}(p,r)$

We now come to the main object of interest in this book: the coclass graph. For a nilpotent group G we write $\gamma(G) = \gamma_{\mathrm{cl}(G)}(G)$ for the last non-trivial term in its lower central series.

Definition 1.22 The *coclass graph* $\mathcal{G}(p,r)$ is a graph whose vertices correspond one-to-one to isomorphism type representatives of finite p-groups of coclass r and there is an edge $G \to H$ if $H/\gamma(H) \cong G$.

If $G \to H$ is an edge in a coclass graph, then $\mathrm{cc}(H) = \mathrm{cc}(G)$ implies that $|H| = p|G|$; thus the order of the groups grows by a factor of p along the edges. If there is a path from H to K in a coclass graph $\mathcal{G}(p,r)$, then H is an *ancestor* of K, and K is a *descendant* of H. We allow paths of length 0, so each group is a descendant of itself. If K has distance k from H, then H is the *k-step ancestor* of K, and K is a *k-step descendant* of H. The 1-step ancestor G of H is the *immediate ancestor* (or *parent*) of H, and H is an *immediate descendant* of G. A group is *terminal* if it has no immediate descendants and it is *non-terminal* (or *capable*) otherwise.

Definition 1.23 Let G be a group in $\mathcal{G}(p,r)$.

a) The *descendant tree* $\Delta(G)$ is the full subtree of $\mathcal{G}(p,r)$ consisting of all descendants of G in $\mathcal{G}(p,r)$.
b) The tree $\Delta(G)$ is a *coclass tree* if it has a unique maximal infinite path. A coclass tree is *maximal* if it is not properly contained in any other coclass tree.
c) If $\Delta(G)$ is a coclass tree, then its unique maximal infinite path $S_1 \to S_2 \to \ldots$ is its *mainline*. The subtree $\mathcal{B}_i$ in $\Delta(G)$ of all descendants of S_i that are not descendants of S_{i+1} is its *i-th branch*.

By definition, each coclass tree of $\mathcal{G}(p,r)$ consists of its mainline and a sequence of branches, and each branch is a finite tree whose root lies on the mainline of the coclass tree. This indicates that the investigation of a coclass tree essentially reduces to an investigation of its branches. The importance of the mainline in a coclass tree is underlined by the following proposition whose proof uses Coclass Theorem D.

Proposition 1.24 *There is a one-to-one correspondence among*

(1) *the maximal infinite paths in* $\mathcal{G}(p,r)$,

(2) *the maximal coclass trees in* $\mathcal{G}(p,r)$, *and*

(3) *the isomorphism types of infinite pro-p-groups of coclass r.*

Proof We first establish the correspondence between (1) and (3). Let G be an infinite pro-p-group of coclass r and write $G_j = G/\gamma_j(G)$ for $j \geqslant 2$. Proposition 1.20 shows that there is $k \geqslant 2$ such that the path $G_k \to G_{k+1} \to \dots$ lies in $\mathcal{G}(p,r)$. If we assume that k is chosen as small as possible with this property, then this infinite path is maximal in $\mathcal{G}(p,r)$: If it can be extended to the left, say $H \to G_k \to \dots$, then $H = G_k/\gamma(G_k) \cong (G/\gamma_k(G))/(\gamma_{k-1}(G)/\gamma_k(G)) \cong G/\gamma_{k-1}(G)$ is a finite p-group of coclass r, contradicting the minimality of k. If H is a pro-p-group isomorphic to G, then each $H/\gamma_j(H) \cong G/\gamma_j(G)$, so H defines the same maximal infinite path as G. Conversely, let $S_1 \to S_2 \to \dots$ be a maximal infinite path in $\mathcal{G}(p,r)$ and for each $i \geqslant 1$ choose an epimorphism $\pi_i \colon S_{i+1} \to S_i$. By definition, the inverse limit G of this linear inverse system is an infinite pro-p-group of coclass r. The proof of Proposition 1.20 shows that for each S_i there is some index $j(i)$ such that $S_i \cong G_{j(i)}$ where $G_k = G/\gamma_k(G)$ as above. Thus, the path $S_1 \to S_2 \to \dots$ lies on the maximal infinite path $G_k \to G_{k+1} \to$ defined by G. Since both paths are maximal infinite in $\mathcal{G}(p,r)$, they coincide. It remains to show that if two infinite pro-p-groups G and H define the same maximal infinite path, then $G \cong H$: this follows from Propositions 1.19 and 1.20 which show that both G and H are isomorphic to the inverse limit S of linear inverse systems with groups $G/\gamma_k(G) \cong H/\gamma_k(H)$, and the isomorphism type of S is independent of the chosen epimorphisms. This completes the proof of the correspondence between (1) and (3).

We now consider the correspondence between (1) and (2). Each maximal coclass tree in $\mathcal{G}(p,r)$ has a unique mainline and this extends to a unique maximal infinite path in $\mathcal{G}(p,r)$. Different maximal coclass trees yield different maximal infinite paths. Conversely, consider a maximal infinite path $K_1 \to K_2 \to \dots$ in $\mathcal{G}(p,r)$. We claim that there exists some K_i so that $\Delta(K_i)$ has a unique infinite path with root K_i and thus is a coclass tree. Suppose there exists no such K_i. Then K_1 is the root of infinitely many maximal infinite paths in $\mathcal{G}(p,r)$, and the correspondence between (1) and (3) implies that there are infinitely many isomorphism types of infinite pro-p-groups of coclass r. This contradicts Coclass Theorem D. Thus, there exists $i \geqslant 1$ such that $\Delta(K_i)$ is a coclass tree; if we choose i as small as possible with this property, then $\Delta(K_i)$ is a maximal coclass tree. This completes the proof. □

Every coclass graph has at least one maximal infinite path as we show in the following example. If the group G acts on an abelian group M, then M is a *G-module*, and the associated *semidirect product* $G \ltimes M$ is the set $\{(g,m) \mid g \in G, m \in M\}$

with the multiplication $(g, m)(h, n) = (gh, m^h n)$, where m^h denotes the image of m under the action of h.

Example 1.25 Consider a prime p, integer $r \geqslant 1$, and write $d = (p-1)p^{r-1}$. Let c be the companion matrix of the polynomial $1 + \mathtt{x}^{p^{r-1}} + \ldots + \mathtt{x}^{(p-2)p^{r-1}} + \mathtt{x}^d$; this is the p^r-th cyclotomic polynomial. Then $P = \langle c \rangle$ is a subgroup of $\mathrm{GL}_d(\mathbb{Z})$ of order p^r that acts naturally on the abelian group $T = (\mathbb{Z}^d, +)$ via matrix multiplication. We define $S = S_{\mathbb{Z}}(p^r)$ as the semidirect product $S = P \ltimes T$. A direct calculation shows that the lower central series quotients $S_i = S/\gamma_{i+1}(S)$ for $i \in \mathbb{N}$ form an infinite path in $\mathcal{G}(p, r)$. We refer to Proposition 3.16 for a proof.

A variation of Example 1.25 arises later in different contexts. First, it is often considered over the p-adic integers $\mathbb{Z}_p$ instead of the integers $\mathbb{Z}$. This makes no difference for the discussion so far, but it will be important later. Second, it translates to number theory over the p-adic integers by viewing c as a p^r-th root of unity over the p-adic rationals $\mathbb{Q}_p$; this point of view will be of interest later and Chapter 3 will reconsider these aspects. An immediate consequence of Example 1.25 is the following observation.

Proposition 1.26 *Every coclass graph $\mathcal{G}(p, r)$ has an infinite path whose root is elementary abelian of rank $r + 1$.*

Proof Let $S = S_{\mathbb{Z}}(p)$ be the group defined in Example 1.25 for $r = 1$ and let E be the elementary abelian p-group of rank $r - 1$. Now consider the direct product $D = S \times E$. Its lower central series quotients are $H_i = D/\gamma_i(D)$. If $i \geqslant 2$, then $H_i \cong S/\gamma_i(S) \times E$ has coclass r. Hence $H_2 \to H_3 \to \ldots$ is a maximal infinite path in $\mathcal{G}(p, r)$ with elementary abelian root H_2. □

The next theorem is the first fundamental result about $\mathcal{G}(p, r)$. Its proof depends on Coclass Theorem D and a deep result [57, Corollary 11.1.5] related to properties of *settled* p-groups.

Theorem 1.27 *Every coclass graph $\mathcal{G}(p, r)$ consists of finitely many maximal coclass trees, and finitely many groups not contained in any of these trees.*

Proof By Proposition 1.24 and Coclass Theorem D each graph $\mathcal{G}(p, r)$ contains only finitely many maximal coclass trees. It remains to show that there are only finitely many groups in $\mathcal{G}(p, r)$ not contained in any of these trees.

Let $\mathcal{M}$ be an infinite set of non-isomorphic finite p-groups of coclass r. Our first aim is to construct infinite subsets $\mathcal{M} = \mathcal{M}_1 \supseteq \mathcal{M}_2 \supseteq \ldots$ and groups $H_1, H_2, \ldots$ such that every $G \in \mathcal{M}_i$ satisfies $G/\gamma_i(G) \cong H_i$. We initialise $\mathcal{M}_1 = \mathcal{M}$ and $H_1 = \{1\}$. Now suppose $\mathcal{M}_i$ and H_i are constructed. The groups $G/\gamma_{i+1}(G)$ with $G \in \mathcal{M}_i$ have order at most p^{i+r} and thus fall into finitely many isomorphism classes. We choose H_{i+1} so that infinitely many $G \in \mathcal{M}_i$ satisfy $G/\gamma_{i+1}(G) \cong H_{i+1}$. We now define $\mathcal{M}_{i+1}$ as the set of groups $G \in \mathcal{M}_i$ with $G/\gamma_{i+1}(G) \cong H_{i+1}$.

Thus every infinite set $\mathcal{M}$ as defined above allows us to define a sequence of finite p-groups $H_1, H_2, \ldots$ such that $H_i \cong H_{i+1}/\gamma_i(H_{i+1})$ for all $i \in \mathbb{N}$. It is shown in [57, Corollary 11.1.5] that there exists a constant $c = c(p, r)$ such every finite p-group G of coclass r and class at least c satisfies $\mathrm{cc}(G) = \mathrm{cc}(G/\gamma_{c+1}(G))$. This implies that $\mathrm{cc}(H_i) = r$ for all $i > c$. In turn, this yields an infinite path $H_{c+1} \to H_{c+2} \to \ldots$ in $\mathcal{G}(p, r)$. This path embeds into a maximal infinite path and therefore corresponds to a maximal coclass tree by Proposition 1.24.

In summary, every infinite set of groups in $\mathcal{G}(p, r)$ contains groups that are part of a coclass tree. This implies that the set of groups that are not contained in any coclass tree of $\mathcal{G}(p, r)$ is finite, as claimed. □

Theorem 1.27 is illustrated in Figure 1.3: The graph $\mathcal{G}(p, r)$ is partitioned into finitely many groups and finitely many maximal coclass trees $\mathcal{T}_1, \ldots, \mathcal{T}_m$, each tree $\mathcal{T}_i$ having a unique maximal infinite path (mainline) and branches $\mathcal{B}_{i,1}, \mathcal{B}_{i,2}, \ldots$. When visualising (parts of) a coclass graph, we use the convention that groups of the same order are drawn on the same horizontal level, and that these levels are arranged such that larger groups are below smaller groups. Since this ordering implicitly defines the direction of an edge (namely, from top to bottom), we can draw them as undirected edges. Based on the partition illustrated in Figure 1.3, in later parts of the book we mainly focus on the investigation of maximal coclass trees and, in particular, their branches.

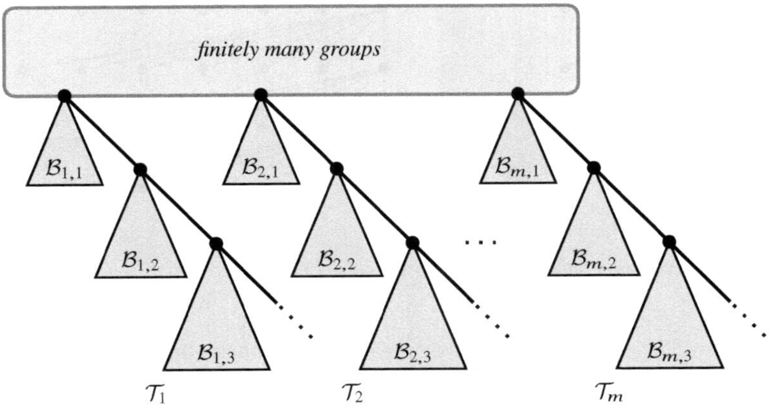

Fig. 1.3 The general shape of a coclass graph $\mathcal{G}(p, r)$

1.6 The examples $\mathcal{G}(2, 1)$ and $\mathcal{G}(3, 1)$

The 2-groups of coclass 1 and the 3-groups of coclass 1 have been fully classified, see Equation (1.1) and the surrounding discussion. This classification can be used readily to determine the coclass graphs $\mathcal{G}(2, 1)$ and $\mathcal{G}(3, 1)$. These graphs are both infinite, as also shown in Proposition 1.26, and we exhibit finite parts of them in Figure 1.4 below.

Both graphs consist of an isolated point corresponding to the cyclic group C_{p^2} and a maximal coclass tree whose root is the elementary abelian group of order p^2. Each of these maximal coclass trees has a mainline. In the case of $\mathcal{G}(2, 1)$, the groups on the mainline are the dihedral groups D_{2^i} for $i \geqslant 2$. These groups can also be described as $S_{\mathbb{Z}}(2)/\gamma_i(S_{\mathbb{Z}}(2))$ for $i \geqslant 2$, where $S_{\mathbb{Z}}(2)$ is the group in Example 1.25. In the case of $\mathcal{G}(3, 1)$, the groups on the mainline are $S_{\mathbb{Z}}(3)/\gamma_i(S_{\mathbb{Z}}(3))$ for $i \geqslant 2$.

In both graphs, the branches of the maximal coclass tree exhibit a periodic pattern. Let $\mathcal{B}_i$ be the branch with root D_{2^i} of order 2^{i+1} in the maximal coclass tree of $\mathcal{G}(2, 1)$. Then there are graph isomorphisms $\mathcal{B}_i \cong \mathcal{B}_{i+1}$ for each $i \geqslant 2$. Similarly, if $\mathcal{B}_i$ is the branch with root of order 3^{i+1} in the maximal coclass tree of $\mathcal{G}(3, 1)$, then $\mathcal{B}_i \cong \mathcal{B}_{i+2}$ holds for all $i \geqslant 3$.

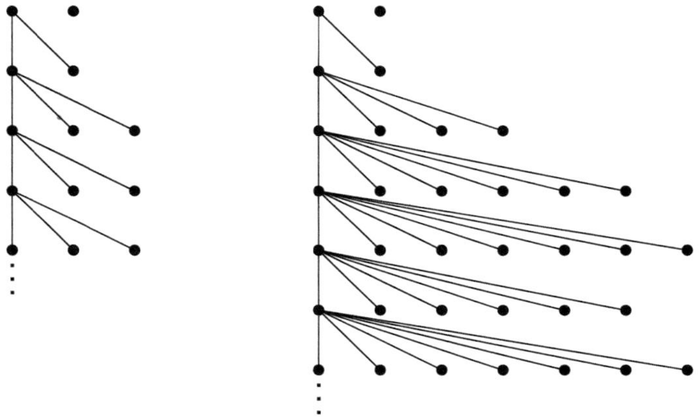

Fig. 1.4 The coclass graphs $\mathcal{G}(2, 1)$ (left) and $\mathcal{G}(3, 1)$ (right)

Thus a full description of each of the infinite graphs $\mathcal{G}(2, 1)$ and $\mathcal{G}(3, 1)$ is obtained from the graph isomorphism types of its branches and a description of the periodic pattern of the branches; see Figures 1.5 and 1.6.

Moreover, the graph isomorphisms between branches can be chosen such that they partition the infinitely many groups into finitely many infinite families. Each of the families corresponds to one of the parametrised presentations in (1.1). These illustrations of $\mathcal{G}(2, 1)$ and $\mathcal{G}(3, 1)$ will form a guiding example in later investigations of general coclass graphs.

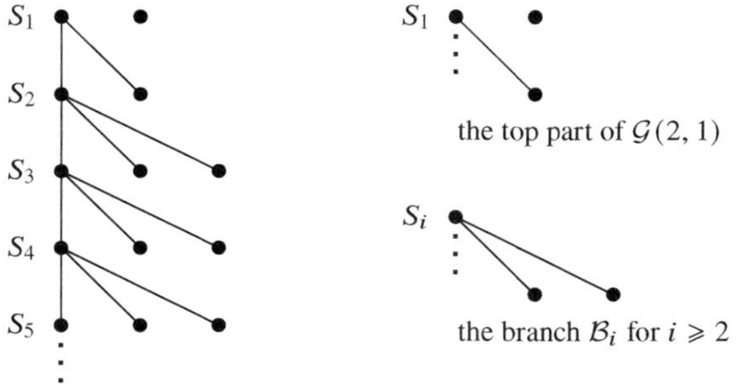

Fig. 1.5 Describing $\mathcal{G}(2, 1)$ (left) via a finite subgraph and a periodic pattern with periodicity 1 (right): $\mathcal{B}_i \cong \mathcal{B}_{i+1}$ for $i \geqslant 2$, and all these branches are connected via $S_1 \rightarrow S_2 \rightarrow \dots$.

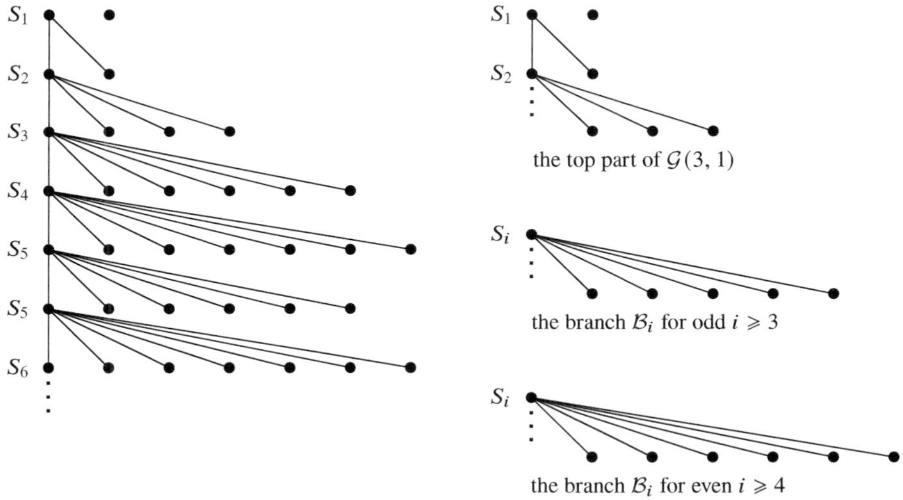

Fig. 1.6 Describing $\mathcal{G}(3, 1)$ (left) via a finite subgraph and a periodic pattern with periodicity 2 (right): $\mathcal{B}_i \cong \mathcal{B}_{i+2}$ for $i \geqslant 3$, and all these branches are connected via $S_1 \rightarrow S_2 \rightarrow \dots$.

Chapter 2
Constructive methods and visualisations

Computational methods have been used extensively to investigate coclass graphs. The well-known *p-group generation* algorithm [68] can be used to compute finite parts of the coclass graphs $\mathcal{G}(p, 1)$. Here we introduce a variation of this algorithm that allows the determination of finite parts of $\mathcal{G}(p, r)$ for each coclass $r \in \mathbb{N}$. We discuss various applications of this algorithm.

2.1 Algorithmic methods

Recall the definition of the coclass graph $\mathcal{G}(p, r)$: the vertices correspond to the isomorphism types of finite p-groups of coclass r and there is an edge $G \longrightarrow H$ if $H/\gamma(H) \cong G$, where $\gamma(H)$ is the last non-trivial term in the lower central series of H. The groups H with $G \longrightarrow H$ are the *immediate descendants* of G. A first step in investigating $\mathcal{G}(p, r)$ is therefore the determination of all immediate descendants of a given group.

2.1.1 Computing immediate descendants

We describe a method that takes as input a finite p-group G and determines a complete and irredundant set of isomorphism type representatives of immediate descendants of G; that is, of the groups H with $\mathrm{cc}(H) = \mathrm{cc}(G)$ and $H/\gamma(H) \cong G$.

Definition 2.1 Let G be a finite p-group of rank d and write $G = F/R$, where F is free of rank d and R is a suitable normal subgroup of F. The *p-multiplier* of G is $M = R/R^*$ and the *p-cover* of G is $G^* = F/R^*$, where $R^* = [F, R]R^{[p]}$.

A group E is an *extension* of a group G by a group M if there is a normal subgroup N of E such that $N \cong M$ and $E/N \cong G$. The extension E is a *central extension* if N is central in E. It is a *split extension* if N has a complement in E, that is, if there

H. Dietrich and B. Eick, *Coclass graphs of p-groups*, Springer Monographs in Mathematics, https://doi.org/10.1007/978-3-032-17506-9_2

exists $K \leqslant E$ such that $E = KN$ and $K \cap N = \{1\}$; note that $K \cong G$ in this case and G is a semidirect product of N and K. The latter is sometimes written $E \cong K \ltimes N$ with multiplication $(k, n)(h, m) = (kh, n^h m)$.

By construction, the p-cover G^* of a finite p-group G is an extension of G by the p-multiplier M. We consider this extension in more detail in the following.

Proposition 2.2 *Let G be a non-trivial finite p-group of rank d with p-cover G^* and p-multiplier M.*

a) *The p-multiplier is a finite elementary abelian subgroup of $\Phi(G^*) \cap Z(G^*)$.*

b) *The p-cover G^* is a finite p-group of rank d with $G^*/M \cong G$.*

c) *Let H be a finite p-group of rank d. If H has a central elementary abelian subgroup N with $H/N \cong G$, then $H \cong G^*/U$ for some $U \leqslant M$.*

d) *The isomorphism type of G^* is independent of the choice of R.*

Proof a) Since R has index $|G|$ in the free group F of rank d, the Nielsen-Schreier Theorem [72, (6.1.1)] shows that R is free of rank $|G|(d-1)+1$. Hence M is an elementary abelian p-group of rank at most $|G|(d-1)+1$, and so finite. As $M = R/R^*$ and $R^* = [F, R]R^{[p]}$, it follows that M is central in G^*. Since G has rank d, it follows with Proposition 1.7 that $C_p^d \cong G/\Phi(G) \cong F/[F, F]F^{[p]}R$. Since F is free of rank d, we also have $F/[F, F]F^{[p]} \cong C_p^d$. This forces $R \leqslant [F, F]F^{[p]}$ and, as a consequence, $M = R/R^*$ is contained in $\Phi(G^*) = [F, F]F^{[p]}/R^*$.

b) By a), the group G^* is a central extension of G by M. Thus $|G^*| = |G||M|$ is finite. As $M \leqslant \Phi(G^*)$, Proposition 1.6 shows that the ranks of G and G^* coincide.

c) Let $\psi\colon F \to G$ be the natural epimorphism with kernel R, and let $\{f_1, \ldots, f_d\}$ be a free generating set for F. Let $\varphi\colon H \to G$ be an epimorphism with central elementary abelian kernel N. For each $i \in \{1, \ldots, d\}$ define $g_i = f_i^{\psi}$ and choose $h_i \in H$ with $h_i^{\varphi} = f_i^{\psi} = g_i$. Now define $\gamma\colon F \to H$ via $f_i^{\gamma} = h_i$ for $i \in \{1, \ldots, d\}$. As F is free on $\{f_1, \ldots, f_d\}$, it follows that γ is a homomorphism with $\psi = \gamma\varphi$. We now show that γ induces the claimed isomorphism $G^*/U \cong H$ for a suitable $U \leqslant M$. First, we establish that γ is onto. Since $G \cong H/N$, Proposition 1.7 implies that $G/\Phi(G) \cong H/\Phi(H)N$, and now $N \leqslant \Phi(H)$ follows from our assumption that G and H both have rank d, that is, $G/\Phi(G) \cong H/\Phi(H) \cong C_p^d$. Note that $\gamma\varphi$ is an epimorphism, so the image I of γ satisfies $IN/N \cong G$, hence $H = IN$. But the elements of N are non-generators by Proposition 1.6, so $H = I$ and γ is onto. Since $\psi = \gamma\varphi$, the kernel K of γ lies in the kernel R of ψ. Thus, $\beta\colon F/K \to G$, $f_iK \to g_i$, is a well-defined epimorphism. If we identify $F/K \cong H$ via $f_iK \to h_i$, then β corresponds to the given epimorphism $\varphi\colon H \to G$, $h_i \to g_i$, with central elementary abelian kernel. It follows that the kernel R/K of β is central elementary abelian. This implies that $R^* \leqslant K$, and allows us to define $U = K/R^*$. Now observe that $U \leqslant M$ and $G^*/U \cong F/K \cong H$, as claimed.

d) Let R and T be normal subgroups of F such that $F/R \cong G \cong F/T$; we show that $F/R^* \cong F/T^*$. Part b) proves that F/R^* and F/T^* have the same rank as G, and both are central extensions of G by an elementary abelian subgroup. Part c) implies that F/T^* is a quotient of F/R^*, and vice versa. As F/R^* and F/T^* are finite, they are isomorphic. □

Proposition 2.2 leads to the following characterisation of the immediate descendants of a finite p-group G; that is, the groups H with $\mathrm{cc}(H) = \mathrm{cc}(G)$ and $H/\gamma(H) \cong G$.

Proposition 2.3 *Let G be a non-trivial finite p-group of class c, let G^* be the p-cover of G with p-multiplier M and let $N = \gamma_{c+1}(G^*)$.*

a) *If H is an immediate descendant of G, then there exists a maximal subgroup U in M with $M = NU$ and $H \cong G^*/U$.*

b) *If U is a maximal subgroup of M with $M = NU$, then G^*/U is an immediate descendant of G.*

Proof As preliminary step in the proof note that an immediate descendant H of G satisfies $|H| = |G|p$ and thus $|\gamma(H)| = p$, since $\mathrm{cc}(G) = \mathrm{cc}(H)$ holds.

a) Let H be an immediate descendant of G. As $\gamma(H) \leqslant \gamma_2(H) \leqslant [H,H]H^{[p]} = \Phi(H)$, it follows that $\Phi(G) \cong \Phi(H)/\gamma(H)$, and so $G/\Phi(G) \cong H/\Phi(H)$. Hence H and G have the same rank. Further, H is an extension of G by the central subgroup $\gamma(H)$ and $\gamma(H)$ has order p. Thus, we can apply Proposition 2.2c) which shows that $H \cong G^*/U$ for some subgroup U of the p-multiplier M of G with $[M : U] = p$. The subgroup U cannot contain $\gamma_{c+1}(G^*)$, since $\mathrm{cl}(H) = \mathrm{cl}(G) + 1 = c + 1$. Hence $UN = M$.

b) The group G^*/U is an extension of $G \cong G^*/M$ by the central subgroup M/U of order p. Further, $\gamma_{c+1}(G^*/U) = \gamma_{c+1}(G^*)U/U = NU/U = M/U$. Thus $\mathrm{cl}(G^*/U) = \mathrm{cl}(G) + 1 = c + 1$ and G^*/U is an immediate descendant of G. □

Proposition 2.3 provides a strategy to construct a complete list of immediate descendants of a non-trivial finite p-group G. It remains to solve the isomorphism problem for them. For this purpose, define

$$\mathrm{Aut}_M(G^*) = \{\alpha \in \mathrm{Aut}(G^*) \mid M^\alpha = M\}$$

the setwise stabiliser of M in $\mathrm{Aut}(G^*)$, and

$$\varrho\colon \mathrm{Aut}_M(G^*) \to \mathrm{Aut}(G), \qquad \alpha \mapsto \alpha_G,$$

where we identify $G = G^*/M$ and define α_G as the automorphism that maps gM to $g^\alpha M$. We investigate the homomorphism ϱ in the following.

Proposition 2.4 *Let G be a non-trivial finite p-group with p-cover G^* and p-multiplier M. Then* $\mathrm{im}(\varrho) = \mathrm{Aut}(G)$*, and* $\ker(\varrho)$ *is an elementary abelian p-group whose elements act trivially on M.*

Proof We first consider the kernel of ϱ. Let $\alpha \in \ker(\varrho)$. For each $g \in G^*$ there exists $m_g \in M$ with $g^\alpha = gm_g$. Since M is central and elementary abelian, if $g, h \in G^*$, then $[g,h]^\alpha = [g^\alpha, h^\alpha] = [gm_g, hm_h] = [g,h]$ and $(g^p)^\alpha = (g^\alpha)^p = (gm_g)^p = g^p$, so α induces the identity on $\Phi(G^*) = [G^*, G^*](G^*)^{[p]}$, and hence on M as $M \leqslant \Phi(G^*)$ by Proposition 2.2. It follows that the elements of $\ker(\varrho)$ commute and have order dividing p. Thus $\ker(\varrho)$ is elementary abelian, as claimed.

Next, we show that ϱ is onto. As before, we identify $G = F/R$, $G^* = F/R^*$, and $M = R/R^*$ where F is free on $\{f_1, \ldots, f_d\}$ and d is the rank of G. Note that $\{f_1R^*, \ldots, f_dR^*\}$ generates G^*. Let $\beta \in \mathrm{Aut}(G)$ and for each f_i choose $x_i \in F$ with $x_iR = (f_iR)^\beta$. Let $\sigma\colon F \to F$ be defined by $f_i \mapsto x_i$. Since β is an automorphism of F/R, it follows that $R^\sigma \leqslant R$. Thus, $(R^*)^\sigma = ([F,R]R^{[p]})^\sigma \leqslant R^*$, and σ induces a homomorphism $\tau\colon F/R^* \to F/R^*$ that fixes R/R^*. In particular, τ induces the automorphism β, and therefore a surjection $F/R^* \to F/R$, $f_iR^* \mapsto x_iR$, with kernel R/R^*. This implies that $F/R^* = (F/R^*)^\tau(R/R^*)$. Proposition 2.2 shows that $R/R^* \leqslant \Phi(F/R^*)$, so $F/R^* = (F/R^*)^\tau(R/R^*) = (F/R^*)^\tau$ by Proposition 1.6. Since F/R^* is finite, τ is an automorphism. By construction, $M^\tau = M$ and τ is a preimage of β under ϱ. This proves that ϱ is onto. □

Let G be a non-trivial finite p-group with p-cover G^* and p-multiplier M. Proposition 2.4 shows that $\varrho\colon \mathrm{Aut}_M(G^*) \to \mathrm{Aut}(G)$ is surjective. Hence for a given $\beta \in \mathrm{Aut}(G)$ we can choose a preimage $\overline{\beta} \in \mathrm{Aut}_M(G^*)$. Proposition 2.4 implies that the restriction $\overline{\beta}_M$ defines an automorphism $M \to M$ that is independent of the choice of the preimage. This allows us to construct a well-defined homomorphism

$$\rho\colon \mathrm{Aut}(G) \to \mathrm{Aut}(M), \quad \beta \mapsto \overline{\beta}_M.$$

Proposition 2.5 *Let G be a non-trivial finite p-group of rank d, with p-cover G^* and p-multiplicator M. Let $U, V \leqslant M$ such that G^*/U and G^*/V are immediate descendants of G.*

a) *Then $G^*/U \cong G^*/V$ if and only if $U^{\beta^\rho} = V$ for some $\beta \in \mathrm{Aut}(G)$.*

b) *Define $\sigma\colon \mathrm{Aut}(G^*/U) \to \mathrm{Aut}(G)$, $\beta \mapsto \beta_{G^*/M}$. Then $\ker(\sigma)$ is elementary abelian of order p^d and $\mathrm{im}(\sigma) = \{\beta \in \mathrm{Aut}(G) \mid U^{\beta^\rho} = U\}$.*

Proof a) If $U^{\beta^\rho} = V$ for some $\beta \in \mathrm{Aut}(G)$, then $\beta^\rho = \tau_M$ for some $\tau \in \mathrm{Aut}_M(G^*)$ with $U^\tau = V$. In particular, τ induces an isomorphism $G^*/U \to G^*/V$, as claimed. Conversely, suppose $\iota\colon G^*/U \to G^*/V$ is an isomorphism. We identify $G = F/R$, $G^* = F/R^*$, and $M = R/R^*$ where F is free on $\{f_1, \ldots, f_d\}$ and $d = \mathrm{rk}(G)$. For each f_i choose $x_i \in F$ with $((f_iR^*)U)^\iota = (x_iR^*)V$, and let $\sigma\colon F \to F$ be the homomorphism that maps each f_i to x_i. Note that ι maps $\gamma(G^*/U) = M/U$ to $\gamma(G^*/V) = M/V$, and so σ induces an automorphism $\beta \in \mathrm{Aut}(F/M)$ of G. Let $\tau \in \mathrm{Aut}_M(F/R^*)$ be a preimage of β under ϱ, which exists by Proposition 2.4. In fact, the construction of τ in Proposition 2.4 shows that we can assume that τ is induced by σ as well. Now let $X, Y \leqslant R$ such that $U = X/R^*$ and $V = Y/R^*$. Since σ arises from ι, we have $X^\sigma \leqslant Y$, and therefore $U^\tau \leqslant V$. As τ is an automorphism of G^* and $|U| = |V|$ is finite, it follows that $V = U^\tau = U^{\beta^\rho}$, as claimed.

b) Let $\{g_1, \ldots, g_d\}$ be a minimal generating set of G^*/U and let $M/U = \langle c \rangle$. Define $\beta_i\colon G^*/U \to G^*/U$ as the map sending $g_i \mapsto g_ic$ and $g_j \mapsto g_j$ for $j \neq i$. Since $M/U \leqslant \Phi(G^*/U)$, it follows that each β_i is an automorphism of G^*/U. By construction, it lies in the kernel of σ. Conversely, each automorphism in the kernel of σ is a product of the automorphisms $\beta_1, \ldots, \beta_d$. Thus $\ker(\sigma) = \langle \beta_1, \ldots, \beta_d \rangle$, which is elementary abelian of order p^d. The proof of a) with $U = V$ shows that the image of σ is given by $\{\beta \in \mathrm{Aut}(G) \mid U^{\beta^\rho} = U\}$. □

In summary, the immediate descendants of a non-trivial finite p-group G of class c can be computed, up to isomorphism, as follows.

(1) Construct the p-cover G^*.

(2) Construct the p-multiplier M and $N = \gamma_{c+1}(G^*)$.

(3) Determine the set $\mathcal{O} = \{U \leqslant M \mid [M : U] = p \text{ and } UN = M\}$.

(4) Determine the action of $\mathrm{Aut}(G)$ on M via Proposition 2.5.

(5) Calculate a set $\mathcal{R}$ of $\mathrm{Aut}(G)$-orbit representatives in $\mathcal{O}$.

(6) Return $\{G^*/U \mid U \in \mathcal{R}\}$.

A practical method for Step (1) is described in [46, Section 9.4.2]. It returns a consistent polycyclic presentation for G^* which permits effective structural computations with the group, see [46, Chapter 8] for details. Based on this, Steps (2) – (6) can be realised directly, and each of the returned groups is also given by a consistent polycyclic presentation. Additionally, the automorphism groups of the resulting groups G^*/U can be determined by Proposition 2.5.

The computation of the maximal subgroups of M in Step (3) and the orbit calculation in Step (5) are usually the bottlenecks that are space and time consuming. One improvement is to use the action of $\mathrm{Aut}(G)$ on the *dual space* of M: We identify the elementary abelian p-multiplier M with a row vector space over the field with p elements. This vector space has a bilinear form $(u, v) = uv^{\intercal}$ for $u, v \in M$, and mapping a maximal subspace U of M to its orthogonal complement

$$U^{\perp} = \{v \in M \mid (v, u) = 0 \text{ for all } u \in U\}$$

yields a bijection between the set of maximal subgroups of M and the set of minimal subgroups. It can be used to translate $\mathcal{O}$ to a set of significantly smaller subgroups (of order p) in M. Various further ideas to facilitate Step (5) rely on effective algorithms to construct the automorphism group of finite p-groups; we refer to the paper [31] for more details.

Example 2.6 We apply the above strategy to

$$G = \langle a, b, c \mid a^4, b^2, c^2, b^a = bc, c^a = c, c^b = c\rangle$$

which is a split extension of C_4 by $C_2 \times C_2$. It has order 16, class 2 and group ID $(16, 3)$ in the SmallGroups Library of the computer algebra systems GAP [40] and MAGMA [3]. The p-cover G^* of G has order 256, class 3 and group ID $(256, 13)$. Hence the multiplier M is elementary abelian of order 2^4. Its subgroup $\gamma_3(G^*)$ has order 4, and therefore M has $2^4 - 1$ maximal subgroups. Among these, there are 3 subgroups containing $\gamma_3(G^*)$. Thus there are $15 - 3 = 12$ maximal subgroups supplementing $\gamma_3(G^*)$. Using orthogonal complements, these subgroups correspond to a set $\mathcal{O}$ of 12 elements of M. The automorphism group $\mathrm{Aut}(G)$ has order 32 and acts as a group of order 4 on M. This group has 6 orbits in $\mathcal{O}$, of lengths $4, 2, 2, 2, 1, 1$. These orbits lead to 6 isomorphism types of immediate descendants; they have group IDs $(32, 6)$, $(32, 7)$, $(32, 8)$, $(32, 9)$, $(32, 10)$, $(32, 11)$ in the SmallGroups library.

2.1.2 The p-group generation algorithm

There are several special cases in which the p-group generation algorithm, see [68] or [46, Section 9.4], provides an alternative method to determine immediate descendants. We now compare the approach of p-group generation with the new method of Section 2.1.1.

Definition 2.7 The *lower exponent-p central series* of a group G is $G = \lambda_1(G) \geqslant \lambda_2(G) \geqslant \ldots$, where $\lambda_{i+1}(G) = [\lambda_i(G), G]\lambda_i(G)^{[p]}$ for each $i \geqslant 0$.

The lower exponent-p central series is the shortest descending central series of G whose quotients are elementary abelian. If G is a finite p-group, then $\lambda_2(G) = \Phi(G) = \gamma_2(G)G^{[p]}$ and, by induction, $\gamma_i(G) \leqslant \lambda_i(G)$ for every $i \geqslant 2$. Analogous to $\gamma(G)$, we define $\lambda(G)$ as the last non-trivial term in the lower exponent-p central series of G. If $\lambda(G) = \lambda_c(G)$, then G has *exponent-p class c*.

Definition 2.8 Let G and H be two finite p-groups. Then H is an immediate *λ-descendant* of G if $H/\lambda(H) \cong G$.

The p-group generation algorithm allows the determination of a complete and irredundant set of isomorphism types of immediate λ-descendants of a finite p-group G. This method can be readily adapted so that it determines only those immediate λ-descendants H with $|H| = p|G|$. However, the groups returned do not necessarily coincide with the immediate descendants G. For example, let

$$H = \langle a, b, c \mid a^4, b^2, c^2, b^a = bc, c^a = c, c^b = c \rangle.$$

Then $\gamma(H) = \langle c \rangle \cong C_2$ and H is an immediate descendant of $G = H/\gamma(H) \cong C_4 \times C_2$, but it is not an immediate λ-descendant of G, because H and G both have 2-class 2. The group G, in turn, is an immediate λ-descendant of $K = G/\lambda(G) \cong C_2^2$, but G is not an immediate descendant of K since the groups have different coclass. The following proposition lists some conditions which imply that immediate descendants and immediate λ-descendants coincide.

Proposition 2.9 *Let G be a finite p-group of coclass r and order p^n. If one of the following conditions holds, then the immediate descendants of G are exactly the immediate λ-descendants of G of order p^{n+1}:*

(1) *$n \geqslant 8p^r + r$;*

(2) *G is non-abelian and $G/\gamma_2(G)$ is elementary abelian;*

(3) *G is non-abelian with rank $r + 1$;*

(4) *G has maximal class and $n \geqslant 3$.*

Proof (1) It follows from [67, Theorem 6.1] that if $n \geqslant 8p^r + 1$, then G has lower exponent-p coclass r. We now prove the claim.
If H is an immediate descendant of G, then $\gamma(H) = \gamma_{c+1}(H)$ is cyclic of order p, and $\gamma_c(G) = \gamma_c(H)/\gamma_{c+1}(H)$ is non-trivial and contained in $\lambda_c(G) = \lambda_c(H)/\gamma_{c+1}(H)$. Since $\lambda_c(G)$ is elementary abelian, this implies that $\lambda_{c+1}(H) \leqslant \gamma_{c+1}(H)$. Note that $\gamma_i(H) \leqslant \lambda_i(H)$ for all i, so $\lambda_{c+1}(H) = \gamma_{c+1}(H)$. We saw above that $\gamma_{c+1}(H) =$

$\gamma(H) \cong C_p$, and therefore $\lambda(H) = \gamma(H)$. This proves that H is an immediate λ-descendant of G of order p^{n+1}.

Conversely, let H be an immediate λ-descendant of G with $|H| = p^{n+1}$. By assumption, H has class and exponent-p class $c+1$, hence $\lambda(H) = \lambda_{c+1}(H)$ has order p and $\gamma_{c+1}(H) \leqslant \lambda_{c+1}(H)$ is non-trivial, hence $\lambda_{c+1}(H) = \gamma_{c+1}(H)$. This shows $\gamma(H) = \lambda(H)$, and H is an immediate descendant of G.

(2) We first prove by induction that if (2) holds, then

$$(*) \quad \gamma_i(G) = \lambda_i(G) \text{ for all } i \geqslant 2;$$

in particular, $\gamma(G) = \lambda(G)$. If $i = 2$, then $\gamma_2(G) = \Phi(G) = \lambda_2(G)$ follows from the assumption and Proposition 1.7. Now assume that $\gamma_j(G) = \lambda_j(G)$ for $j = 2, \ldots, i$. The group $\gamma_i(G)/\gamma_{i+1}(G)$ is generated by all cosets $[s,t]\gamma_{i+1}(G)$ with $s \in G$ and $t \in \gamma_{i-1}(G)$. By the induction hypothesis, $\gamma_{i-1}(G)/\gamma_i(G)$ has exponent p, so $t^p \in \gamma_i(G)$. Commutator calculations show that $[s,t]^p\gamma_{i+1}(G) = [s,t^p]\gamma_{i+1}(G) = \gamma_{i+1}(G)$, and we conclude that $\gamma_i(G)/\gamma_{i+1}(G)$ is elementary abelian. Since $\gamma_i(G) = \lambda_i(G)$ by the induction hypothesis, hence $\gamma_{i+1}(G) = \lambda_{i+1}(G)$, as claimed. This proves $(*)$, and we now show that $(*)$ implies the claim.

If H is an immediate descendant of G, then $\gamma(H) \leqslant \gamma_2(H)$, and $G \cong H/\gamma(H)$ implies that $H/\gamma_2(H) \cong G/\gamma_2(G)$ is elementary abelian as well. Thus, H satisfies Condition (2), and $(*)$ applied to H proves that $\gamma(H) = \lambda(H)$, which has order p. Therefore H is an immediate λ-descendant of G of order p^{n+1}.

Conversely, if H is an immediate λ-descendant of G of order p^{n+1}, then $\lambda(H) = \lambda_{c+1}(H)$ has order p and $H/\lambda(H) \cong G$. Together with $(*)$, this implies that $\gamma_i(H)\lambda(H) = \lambda_i(H)$ for all $i \in \{2, \ldots, c+1\}$. We make a case distinction on whether $\lambda(H)$ is contained in $\gamma_2(H)$. First suppose that $\lambda(H) \nleqslant \gamma_2(H)$. Since $\lambda(H)$ is central of order p, this implies that $\lambda_i(H) = \gamma_i(H) \times \lambda(H)$ for all $i \in \{2, \ldots, c+1\}$. However, $\lambda_2(H) = \gamma_2(H) \times \lambda(H)$ yields $\lambda_3(H) = [H, \lambda_2(H)]\lambda_2(H)^{[p]} \leqslant [H, \gamma_2(H)]\gamma_2(H) \leqslant \gamma_2(H)$, which is a contradiction to $\lambda(H) \nleqslant \gamma_2(H)$, unless $\lambda(H) = \lambda_2(H)$ and $\gamma_2(H) = \{1\}$. But then H is abelian, which contradicts our assumption that G is non-abelian. We conclude that $\lambda(H) \leqslant \gamma_2(H)$. Then $\gamma_2(H) = \lambda_2(H)$, which shows that $H/\gamma_2(H)$ is elementary abelian, so (2) holds for H. Now $(*)$ applied to H proves that $\gamma(H) = \lambda(H)$, which has order p. Thus, H is an immediate descendant of G, as claimed.

(3) By assumption, $p^{r+1} = [G : \Phi(G)] \leqslant [G : \gamma_2(G)]$, and $[G : \gamma_2(G)] \leqslant p^{r+1}$ by Lemma 1.12. Thus, $\gamma_2(G) = \Phi(G)$, which implies that $G/\gamma_2(G)$ is elementary abelian and (2) applies.

(4) If $r = 1$ and $n \geqslant 3$, then G has rank 2 by Lemma 1.12; in particular, such a G is non-abelian. Now (3) proves the claim. □

An implementation of p-group generation is available in MAGMA and in the ANUPQ package [39] for GAP. An implementation of the algorithm in Section 2.1.1 is part of the AutPGrp package [36] for GAP.

2.2 Investigating coclass trees

We have seen in Theorem 1.27 that each coclass graph $\mathcal{G}(p,r)$ consists of finitely many maximal coclass trees and finitely many other groups. The focus of the investigation is now on understanding the maximal coclass trees. By definition, such a tree consists of its mainline $S_1 \to S_2 \to \dots$ and its (infinitely many) branches $\mathcal{B}_1, \mathcal{B}_2, \dots$. Recall that each branch $\mathcal{B}_i$ is a finite tree consisting of all groups in the coclass tree that are descendants of S_i but not of S_{i+1}.

Definition 2.10 Let $\mathcal{T}$ be a coclass tree in $\mathcal{G}(p,r)$ with branches $\mathcal{B}_1, \mathcal{B}_2, \dots$.

a) The tree $\mathcal{T}$ has *bounded depth* if there exists $b \in \mathbb{N}$ such that each branch $\mathcal{B}_i$ has depth at most b, that is, each group in $\mathcal{B}_i$ has distance at most b from the root S_i.

b) The tree $\mathcal{T}$ has *bounded width* if there exists $w \in \mathbb{N}$ such that each branch $\mathcal{B}_i$ has at most w groups of any given order.

The features of bounded depth and bounded width play a major role in the investigation of coclass trees. Coclass trees of bounded depth can often be visualised easily. This becomes more difficult for trees with unbounded depth and bounded width. If depth and width are both unbounded, then a meaningful visualisation is often impossible. In the following two subsections, we will first provide an overview of the historical development of coclass tree investigations, and then introduce notation for compact visualisations of these trees.

2.2.1 Some history

The graphs $\mathcal{G}(2,1)$ and $\mathcal{G}(3,1)$ are determined by Blackburn's classification [1]. Finite parts of the coclass trees in $\mathcal{G}(2,1)$ and in $\mathcal{G}(3,1)$ can also easily be obtained using the methods of Section 2.1. For all other coclass graphs, structural properties were proved only after significant computational investigations, as explained below.

The graphs $\mathcal{G}(p,1)$: The p-groups of coclass 1 were investigated by Leedham-Green & McKay in a sequence of papers [53–56]. In these cases Proposition 2.9 applies and hence p-group generation can be used to explore the relevant coclass trees. Newman [66] used this to exhibit an experimental description of $\mathcal{G}(5,1)$, see also [17]. We include an outline of these graphs in Section 2.4. Based on this, one observes that the coclass tree in $\mathcal{G}(5,1)$ has unbounded depth and bounded width. The work of Leedham-Green & McKay also implies that the coclass tree in $\mathcal{G}(p,1)$ for $p \geqslant 7$ has unbounded depth and width. We include a proof for these features in Chapter 9.

The graphs $\mathcal{G}(2,r)$ and Conjecture P: The graphs $\mathcal{G}(2,r)$ for $r \leqslant 3$ have been investigated in detail by Newman & O'Brien [67]. The results led Newman & O'Brien to suggest Conjecture P. This conjecture is nowadays know as *Periodicity I*; it was proved independently by du Sautoy [22] and by Eick & Leedham-Green [32], with a correction by Couson [12]. Periodicity I and a corrected version of the proof of Eick

& Leedham-Green is presented in Chapter 7. In Section 2.3 we exhibit a finite part of the graph $\mathcal{G}(2,2)$.

The graph $\mathcal{G}(3,2)$ and Conjecture W: Only one other graph has been explicitly investigated: $\mathcal{G}(3,2)$ was studied by Eick, Leedham-Green, Newman & O'Brien [33]. This graph has 16 maximal coclass trees: 12 of bounded depth and 4 of bounded width and unbounded depth. The coclass trees in this graph have been explored using p-group generation (despite the fact that Proposition 2.9 does not directly apply). The broad features of the maximal coclass trees were proved in [33]. These results motivated the authors of [33] to propose *Conjecture W*, which is an attempt to describe general coclass trees using two types of periodic patterns; we provide more details in Section 2.5.

2.2.2 Compact visualisations

Coclass trees of bounded width can often be visualised nicely when a compact way of drawing trees is used. We introduce such a drawing convention in the following.

Convention 2.11 A blue number m right of a vertex in a coclass graph indicates that the group represented by this vertex has m terminal immediate descendants. If a vertex has no blue number, then the corresponding group has no terminal immediate descendants. An orange number k left of a vertex G in a coclass graph indicates that the whole descendant tree $\Delta(G)$ of G appears k times such that the root of each of these descendant trees is connected to the parent of G. These conventions are visualised in the example in Figure 2.1.

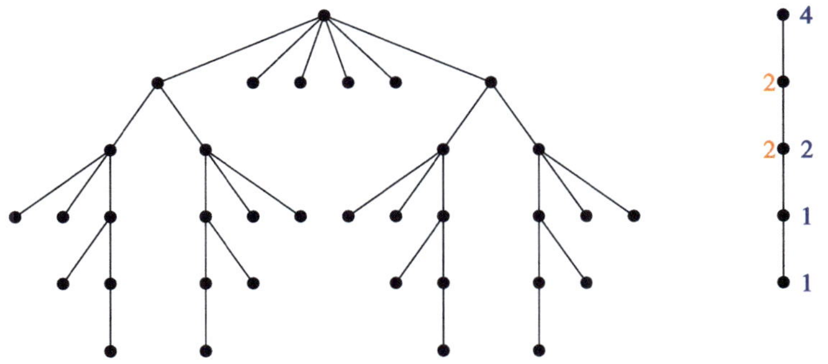

Fig. 2.1 Illustrating Convention 2.11: the right graph depicts the compact form of the graph on the left. Orange numbers (left of a vertex) indicate branchings into isomorphic subtrees; blue numbers (right of a vertex) indicate terminal descendants.

2.3 The graph $\mathcal{G}(2,2)$

The 2-groups of coclass 2 were determined, up to isomorphism, by James [73, 74], with a correction by Newman & O'Brien [67, p. 132]. Based on these classifications, the graph $\mathcal{G}(2,2)$ is well understood. The following result shows how the structure of $\mathcal{G}(2,2)$ can be investigated by computing suitable descendant trees.

Proposition 2.12 *The graph $\mathcal{G}(2,2)$ consists of the cyclic group C_8 and the two descendant trees $\Delta(C_2^3)$ and $\Delta(C_4 \times C_2)$.*

Proof Let G be a finite 2-group of coclass 2. Lemma 1.12 shows that $\mathrm{rk}(G) \leqslant 3$ and $|G/\gamma_2(G)| \leqslant 8$. If $\mathrm{rk}(G) = 1$, then G is cyclic, and hence $G \cong C_8$ is a terminal group in $\mathcal{G}(2,2)$. If $\mathrm{rk}(G) = 3$, then $G/\gamma_2(G) \cong C_2^3$, and hence G lies in $\Delta(C_2^3)$. If $\mathrm{rk}(G) = 2$, then $G/\gamma_2(G)$ is an abelian group of rank 2 and order at most 8. Thus $G/\gamma_2(G) \cong C_2^2$ or $G/\gamma_2(G) \cong C_4 \times C_2$ are the two possible options. In the first case, Blackburn [1, Corollary p. 74] proved that G has coclass 1, thus this case does not arise. In the second case, G lies in $\Delta(C_4 \times C_2)$, which completes the proof. □

Figure 2.2 shows the top part of the coclass graph $\mathcal{G}(2,2)$ displayed in the compact form using Convention 2.11. The groups in the first horizontal layer of Figure 2.2 are, from left to right, the group $C_4 \times C_2$, the elementary abelian group C_2^3, and the cyclic group C_8.

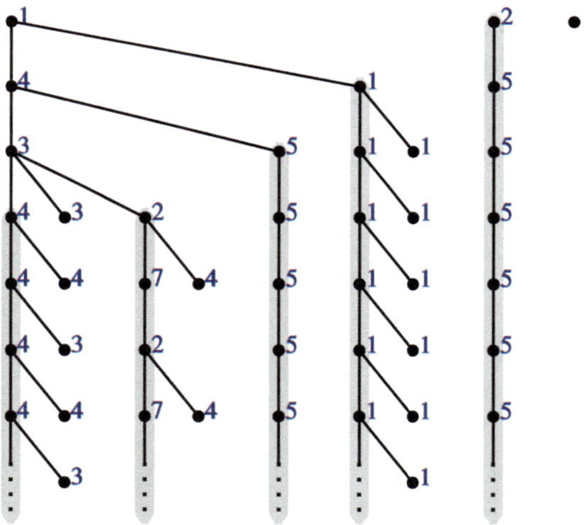

Fig. 2.2 The initial part of $\mathcal{G}(2,2)$ in compact notation, using Convention 2.11. Highlighted in grey are the mainlines of the coclass trees.

We summarise the features of $\mathcal{G}(2,2)$ in the following result.

Proposition 2.13 (Newman & O'Brien [67]) *The coclass graph* $\mathcal{G}(2,2)$ *has five maximal coclass trees and* 16 *groups outside these coclass trees. Each coclass tree has bounded depth: the depths of the branches are either* 1 *or* 2*. Each coclass tree has a virtually periodic pattern: with finitely many exceptions, its branches satisfy* $\mathcal{B}_i \cong \mathcal{B}_{i+1}$ *or* $\mathcal{B}_i \cong \mathcal{B}_{i+2}$.

2.4 A conjectural description of $\mathcal{G}(5, 1)$

While the graphs $\mathcal{G}(2, 1)$ and $\mathcal{G}(3, 1)$ are fully determined, the classification of the 5-groups of coclass 1 is a long-standing open problem. It is especially intriguing, because experimentally we obtain a very clear insight into the precise structure of $\mathcal{G}(5, 1)$. However, it remains difficult to complete the proof of the conjectured classification. Before we consider $\mathcal{G}(5, 1)$, we first give an initial description of $\mathcal{G}(p, 1)$ for arbitrary primes p.

Proposition 2.14 *For each prime p, the graph* $\mathcal{G}(p, 1)$ *consists of the cyclic group* C_{p^2} *and a single descendant tree* $\Delta(C_p^2)$.

Proof If G is a p-group of coclass 1, then $G/\gamma_2(G)$ has rank 1 or 2, and size p^2 by Lemma 1.12. Thus, $G/\gamma_2(G)$ is either cyclic or elementary abelian of order p^2. In the first case, G is cyclic, so $G \cong C_{p^2}$; in the second case, G is contained in $\Delta(C_p^2)$.□

The single descendant tree $\Delta(C_p^2)$ is also a maximal coclass tree. This and other features of $\mathcal{G}(p, 1)$ will be investigated in Chapter 9. Here, we will shift our focus to $\mathcal{G}(5, 1)$ and present the conjectured description of the branches within its coclass tree. To assist the exposition, we use the following notation together with a few features that will be proved in Chapter 9.

Remark 2.15 The tree $\mathcal{T} = \Delta(C_5^2)$ is the unique maximal coclass tree in $\mathcal{G}(5, 1)$. If $\mathcal{B}_i$ is the branch in $\mathcal{T}$ with root of order 5^i, then $\mathcal{B}_i$ has depth i.

Computational methods allow the visualisation of $\mathcal{B}_i$ for small i, see Newman [66] and by Eick, Dietrich & Feichtenschlager [17]. As example, an illustration of the branches $\mathcal{B}_{10}, \mathcal{B}_{14}, \mathcal{B}_{18}$ are displayed in Figure 2.3 using Convention 2.11. Based on such experimental investigations it is easy to suggest two periodic patterns for the branches $\mathcal{B}_i$. The first pattern aligns the top part of $\mathcal{B}_i$ with the top part of $\mathcal{B}_{i+4}$: if $\mathcal{B}_i(k)$ denotes the subtree of $\mathcal{B}_i$ consisting of all groups of depth at most k in $\mathcal{B}_i$, then $\mathcal{B}_i(i-1) \cong \mathcal{B}_{i+4}(i-1)$ seems to hold. The second pattern aligns the bottom part of $\mathcal{B}_{i+4}$ with the bottom part of $\mathcal{B}_i$: the subgraph of the last 5 layers of groups in $\mathcal{B}_{i+4}$ seems to be isomorphic to the corresponding subgraph in $\mathcal{B}_i$. Together, these two conjectured patterns provide a conjectural description of $\mathcal{B}_{i+4}$ in terms of $\mathcal{B}_i$ for all $i \geqslant 10$. We illustrate the resulting conjecture in Figure 2.4.

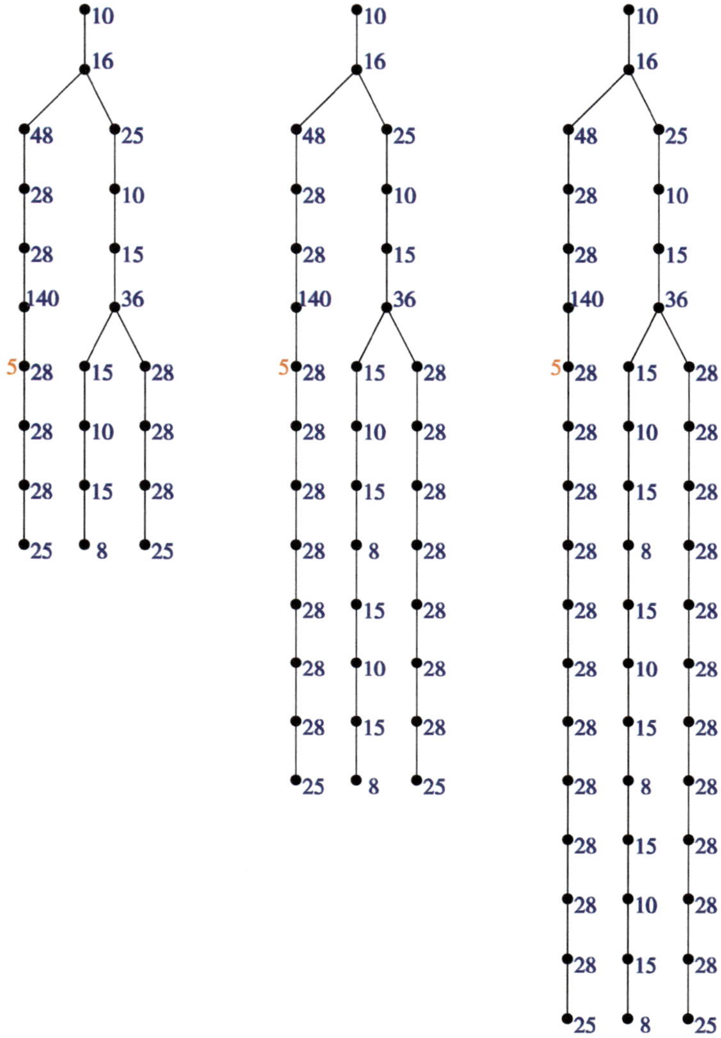

Fig. 2.3 The branches $\mathcal{B}_{10}$, $\mathcal{B}_{14}$, and $\mathcal{B}_{18}$ (left to right) in $\mathcal{G}(5, 1)$

2.5 Towards periodic patterns in coclass graphs

The coclass graphs we have examined thus far reveal intriguing periodic patterns that are central to the idea of classifying finite p-groups by coclass. Consequently, a primary objective of coclass theory is to thoroughly understand the periodic structure of these coclass graphs. Motivated by our discussions of $\mathcal{G}(2, 1)$, $\mathcal{G}(2, 2)$, $\mathcal{G}(3, 1)$, and $\mathcal{G}(5, 1)$, we formulate the following conjecture.

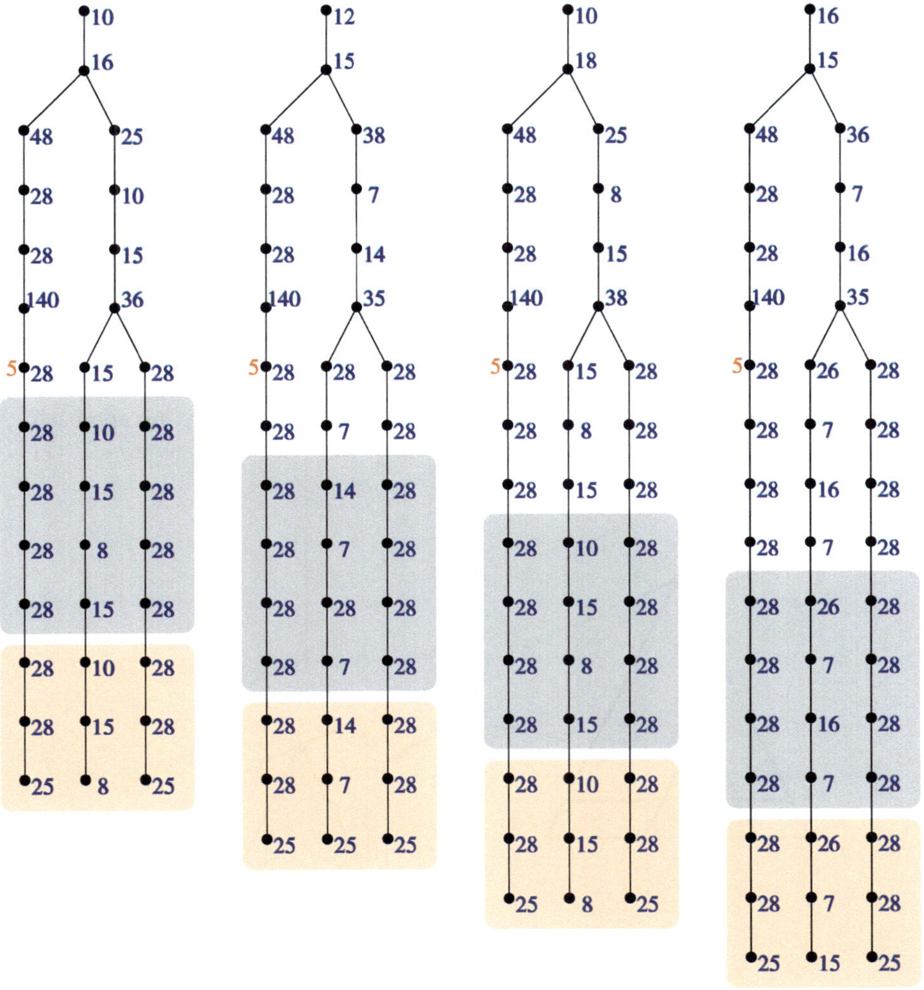

Fig. 2.4 Conjectural description of the branches $\mathcal{B}_{10+4x}, \ldots, \mathcal{B}_{13+4x}$ (left to right) in $\mathcal{G}(5,1)$ for $x \in \mathbb{N}_0$, using Convention 2.11. The grey part is repeated x times so $\mathcal{B}_i$ has depth i. The white and grey parts of these branches are proved, see [13, 14]. The orange parts are conjectured.

Conjecture 2.16 (Coclass Graph Conjecture) Each coclass graph $\mathcal{G}(p,r)$ can be described completely by a finite subgraph and at most two types of periodic patterns.

As we have seen before, for the graphs $\mathcal{G}(2,1)$, $\mathcal{G}(2,2)$, and $\mathcal{G}(3,1)$, one periodic pattern suffices: it considers each maximal coclass tree $\mathcal{T}$ separately and describes the branches of $\mathcal{T}$ for all large enough $i \in \mathbb{N}$ via $\mathcal{B}_i \cong \mathcal{B}_{i+d}$ with $d \in \{1,2\}$ depending on $\mathcal{T}$. We say this pattern (Periodicity I) moves *parallel* to the mainline of $\mathcal{T}$.

Since the coclass tree $\mathcal{T}$ in $\mathcal{G}(5,1)$ has unbounded depth, Periodicity I is not capable of describing $\mathcal{T}$ completely and a second pattern (*Periodicity II*) is necessary.

As before, Periodicity I moves parallel to the mainline of $\mathcal{T}$ and here it observes that $\mathcal{B}_i(i-1) \cong \mathcal{B}_{i+4}(i-1)$ for all large enough i. The second periodicity moves *orthogonal* to the mainline and describes how the bottom part of $\mathcal{B}_{i+4}$ can be recovered from that of $\mathcal{B}_i$.

The conjectured Periodicity II aims to describe the growth of the branches. Supporting evidence for this conjecture stems from the detailed investigations of $\mathcal{G}(5,1)$, the partial descriptions of $\mathcal{G}(p,1)$, and the study of $\mathcal{G}(3,2)$. One formulation of the conjectured Periodicity II is that for every maximal coclass tree $\mathcal{T}$ there exists a periodicity $d \in \mathbb{N}$ such that the following holds: for every group G at *large* depth k in a branch $\mathcal{B}_{i+d}$ there exists a group $P(G)$ at depth $k-d$ in $\mathcal{B}_i$ such that the descendant tree $\Delta(G)$ of G is isomorphic to $\Delta(P(G))$. Figure 2.5 illustrates a version of this conjectured periodicity. For example, $d = 4$ in $\mathcal{G}(5,1)$, and if G has descendants, then $P(G)$ can be constructed by taking the 4-step ancestor of G in $\mathcal{B}_{i+4}$ and then applying the graph isomorphism $\mathcal{B}_{i+4}(i-1) \cong \mathcal{B}_i(i-1)$. Unfortunately, this approach does not generalise and a critical open question is: *How can one determine* $P(G)$ *from* G*?*

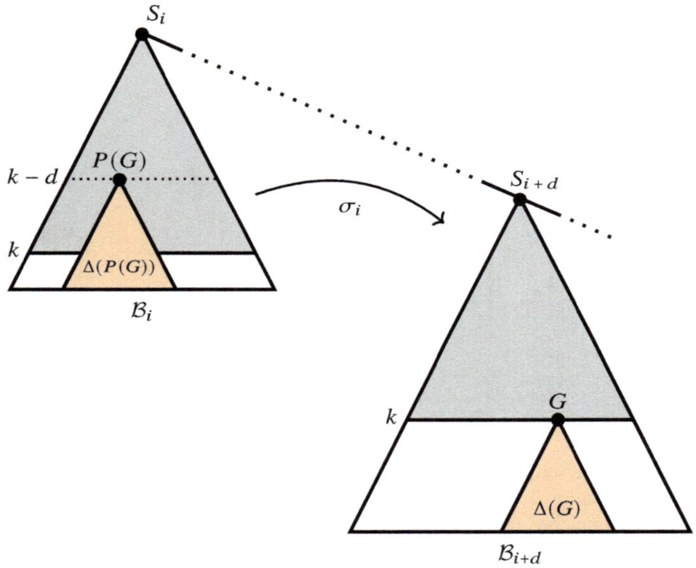

Fig. 2.5 Illustrating Periodicity I and a conjectured version of Periodicity II: Periodicity I observes that the pruned (gray shaded) subtrees $\mathcal{B}_i(k)$ and $\mathcal{B}_{i+d}(k)$ are isomorphic via a graph isomorphism σ_i. Periodicity II conjectures that the descendant tree $\Delta(G)$ of every group G at depth k in $\mathcal{B}_{i+d}$ is isomorphic to the descendant tree $\Delta(P(G))$ of an associated group $P(G)$ at depth $k-d$ in $\mathcal{B}_i$.

Chapter 3
Number theory and group constructions

Number theory plays an important role in coclass theory. We first illustrate this connection with a sample construction in Section 3.1 and then introduce topics required in our later investigations. The results presented in this chapter are known; we provide a unified summary for convenience. For further reading on number theory we refer to the books of Hasse [43] and Neukirch [65], and for p-adic number theory the books of Gouvêa [42] and Robert [71].

3.1 From coclass theory to number theory

Let p be a prime and let $r \geqslant 1$ be an integer. Example 1.25 describes how the companion matrix c of the p^r-th cyclotomic polynomial over $\mathbb{Z}$ determines an infinite group $S_{\mathbb{Z}}(p^r)$ whose lower central series quotients form the mainline of a maximal coclass tree $\mathcal{T}$ in $\mathcal{G}(p,r)$. The group $S_{\mathbb{Z}}(p^r)$ arises as a split extension of $P = \langle c \rangle$ with the natural module $T = (\mathbb{Z}^d, +)$, where $d = p^{r-1}(p-1)$ is the degree of the considered polynomial.

The same construction can be done when $\mathbb{Z}$ is replaced by the p-adic integers $\mathbb{Z}_p$; see Section 3.2 for an introduction to the latter. Specifically, let $T = (\mathbb{Z}_p^d, +)$ and define $S(p^r) = P \ltimes T$ where P is as above. This group is an infinite pro-p-group whose lower central series quotients are isomorphic to those of $S_{\mathbb{Z}}(p^r)$. Thus, $S(p^r)$ also defines the infinite path in $\mathcal{T}$ *and additionally* is isomorphic to the inverse limit of the groups on this path. This is one reason why it is more natural to consider the pro-p-group $S(p^r)$ rather than the integral version. We see later that some inverse limits of infinite paths in coclass trees yield pro-p-groups that have no integral counterpart (see Corollary 4.15), so the use of p-adic groups is necessary.

Leedham-Green & McKay [53–56] give a construction which exploits $S(p^r)$ to construct many additional groups in $\mathcal{T}$. We give a brief outline in the following example (where for now assume the reader is familiar with the construction of exterior squares of abelian groups); for details and proofs see Proposition 3.16 and Chapter 8.

H. Dietrich and B. Eick, *Coclass graphs of p-groups*, Springer Monographs in Mathematics, https://doi.org/10.1007/978-3-032-17506-9_3

Example 3.1 Let $S = S(p^r)$ be the pro-p-group defined above. Write $T_0 = T$ and for $i \in \mathbb{N}$ define $T_i = [T_{i-1}, S] = \gamma_{i+1}(S)$. If $\mathcal{T}$ is the maximal coclass tree in $\mathcal{G}(p, r)$ defined by S, then the mainline groups are S/T_i for $i \in \mathbb{N}$. Consider a P-module homomorphism $f \colon T \wedge T \to T$. The image $\mathrm{im}(f)$ is a P-submodule of T and thus there exists $j \in \mathbb{N}$ with $\mathrm{im}(f) = T_j$. Assume that f is chosen such that $j \geqslant d$ and let $m \in \{j, \ldots, 2j - d\}$. Then f allows us to define the group $G_{f,m}$ with underlying set $\{(g, t + T_m) \mid g \in P, t \in T\}$ and multiplication

$$(g, s + T_m)(h, t + T_m) = (gh, sh + t + \tfrac{1}{2} f(sh \wedge t) + T_m).$$

This group has order p^{r+m} and lies the branch of $\mathcal{T}$ that has root S/T_j. Note that $G_{f,m}$ is a semidirect product of P with a *twisted* version of T/T_m where the multiplication in T/T_m is $(s + T_m)(t + T_m) = s + t + \frac{1}{2} f(s \wedge t) + T_m$.

Following similar ideas, in Chapter 8 we will discuss a more general construction of groups in $\mathcal{G}(p, r)$. In particular, Theorem 8.27 will show that the groups obtained in this way form a subtree of the coclass tree $\mathcal{T}$ that exhibits the *broad structure* of $\mathcal{T}$: for example, we can read off this subtree whether $\mathcal{T}$ has finite depth or width.

The isomorphism problem for the groups in Example 3.1 and the construction of their automorphism groups translate to group actions involving the cyclotomic field $\mathbb{Q}_p(\theta)$, where θ is a p^r-th root of unity. These group actions exploit the Galois group and the ring of integers of this field. The aim of this chapter is to discuss the relevant p-adic number theory.

3.2 Introduction to p-adic numbers

Every nonzero integer a can be written uniquely as $a = p^m b$ where b is an integer coprime to p. With this decomposition, the *p-adic valuation* of a is defined as $\nu_p(a) = m$. This valuation can be extended to nonzero rational numbers $s = a/b$ by defining $\nu_p(s) = \nu_p(a) - \nu_p(b)$. For example,

$$\nu_5(3050/16875) = \nu_5(5^2 \cdot 122) - \nu_5(5^4 \cdot 27) = 2 - 4 = -2.$$

The corresponding *p-adic absolute value* is $|s|_p = p^{-\nu_p(s)}$. It is convenient to extend these definitions to $\nu_p(0) = \infty$ and $|0|_p = 0$. If $s \in \mathbb{Q}$, then $|s|_p \geqslant 0$, and $|s|_p = 0$ if and only if $s = 0$. Moreover, $|\cdot|_p$ is multiplicative, that is, if $s, t \in \mathbb{Q}$, then $|st|_p = |s|_p |t|_p$; it is also non-archimedean, that is,

$$|s + t|_p \leqslant \max\{|s|_p, |t|_p\}.$$

The non-archimedean property has the following consequence.

Lemma 3.2 *If* $s, t \in \mathbb{Q}$ *and* $|s|_p \neq |t|_p$, *then* $|s + t|_p = \max\{|s|_p, |t|_p\}$.

Proof Suppose $|s|_p > |t|_p$, so $|s+t|_p \leqslant |s|_p$. Writing $s = (s+t) - t$, the non-archimedean property implies $|s|_p \leqslant \max\{|s+t|_p, |t|_p\}$, and $|s|_p > |t|_p$ forces $|s|_p \leqslant |s+t|_p$. Thus, $|s+t|_p = |s|_p$, as claimed. □

The absolute value $|\cdot|_p$ makes $\mathbb{Q}$ a metric space with *p-adic metric* $d_p(s,t) = |s-t|_p$. If $s \neq t$ and $s - t = p^n \cdot \frac{a}{b}$ with $p \nmid a, b$, then the larger n is, the *closer* s and t are in this metric.

The p-adic metric allows the definition of Cauchy sequences: these are sequences $(s_n)_{n\in\mathbb{N}}$ of elements of $\mathbb{Q}$ such that for every $\varepsilon > 0$ there exists some $m > 0$ such that $d_p(s_i, s_j) < \varepsilon$ for all $i, j \geqslant m$. Two Cauchy sequences $(s_n)_{n\in\mathbb{N}}$ and $(t_n)_{n\in\mathbb{N}}$ are equivalent if $d_p(s_n, t_n)$ converges to 0 as $n \to \infty$. The set of all equivalence classes forms the well-known completion of the metric space $(\mathbb{Q}, d_p)$, which is the field of p-adic rational numbers. Algebraically, this completion can be constructed as the quotient field of the ring of all Cauchy sequences over $\mathbb{Q}$ modulo the maximal ideal of all Cauchy sequences that are equivalent to 0 (that is, that converge to 0). A field is complete if it is isomorphic to its completion.

Definition 3.3 The field of *p-adic rationals* $\mathbb{Q}_p$ is the completion of $\mathbb{Q}$ with respect to the p-adic metric.

If $(s_n)_{n\in\mathbb{N}}$ is a Cauchy sequence representing a nonzero $s \in \mathbb{Q}_p$, then there exists a constant $c > 0$ and an index $m \in \mathbb{N}$ such that $|s_n|_p \geqslant c > 0$ for all $n \geqslant m$. One can replace m by a larger constant such that also $|s_i - s_j|_p < c$ for all $i, j \geqslant m$. But then $|s_i - s_j|_p < \max\{|s_i|_p, |s_j|_p\}$ for all $i, j \geqslant m$, which forces $|s_i|_p = |s_j|_p$ by Lemma 3.2. This shows that the sequence $(|s_n|_p)_{n\in\mathbb{N}}$ eventually becomes constant, which allows us to extend the p-adic absolute value to $\mathbb{Q}_p$ via $|s|_p = \lim_{n\to\infty} |s_n|_p$. An immediate consequence is the following lemma.

Lemma 3.4 *Every nonzero* $s \in \mathbb{Q}_p$ *can be written uniquely as* $s = p^c u$ *for some* $c \in \mathbb{Z}$ *and* $u \in \mathbb{Q}_p$ *with* $|u|_p = 1$.

The field $\mathbb{Q}_p$ is not algebraically closed, but taking the algebraic closure and constructing its completion $\mathbb{C}_p$ (with respect to the extended p-adic absolute value), there is a non-canonical field isomorphism $\mathbb{C}_p \cong \mathbb{C}$ whose existence depends on the Axiom of Choice, see for example [71, p. 145, Theorem]. Since $\mathbb{Q}_p$ embeds into $\mathbb{C}_p$, the next useful result follows.

Proposition 3.5 *There is a (non-canonical) injective field homomorphism from* $\mathbb{Q}_p$ *into the complex numbers* $\mathbb{C}$.

One consequence of Proposition 3.5 is that every root of unity in $\mathbb{Q}_p$ can be considered as a complex root of unity.

Recall from Example 1.14 that the ring of p-adic integers $\mathbb{Z}_p$ is the inverse limit of the groups $\mathbb{Z}/p^n\mathbb{Z}$, $n = 1, 2, \ldots$. Its elements can be described as sequences $z = (z_n)_{n\in\mathbb{N}}$ with each $z_n \in \{0, 1, \ldots, p^n - 1\}$ such that $z_n \equiv z_m \bmod p^n$ whenever $m \geqslant n$. This shows that z is a Cauchy sequence in $\mathbb{Q}$ with respect to the p-adic metric, and $|z|_p \leqslant 1$ since every component z_n is an integer. The condition that each z_n lies in $\{0, 1, \ldots, p^n - 1\}$ implies that distinct elements of $\mathbb{Z}_p$ define distinct equivalence classes of Cauchy sequences in $\mathbb{Q}_p$; thus we can consider $\mathbb{Z}_p \subseteq \{s \in \mathbb{Q}_p \mid |s|_p \leqslant 1\}$. In fact, the converse holds.

Proposition 3.6 *The* p-adic integers *are* $\mathbb{Z}_p = \{s \in \mathbb{Q}_p \mid |s|_p \leqslant 1\}$.

Proof The inclusion $\mathbb{Z}_p \subseteq \{s \in \mathbb{Q}_p \mid |s|_p \leqslant 1\}$ was shown above. Now let $s \in \mathbb{Q}_p$ be nonzero with $|s|_p \leqslant 1$. By Lemma 3.4, we can write $s = p^c u$ for some $c \geqslant 0$ and $u \in \mathbb{Q}_p$ with $|u|_p = 1$. The claim follows if we show that u is represented by an element of $\mathbb{Z}_p$. Let $t = (t_n)_{n\in\mathbb{N}}$ be a Cauchy sequence in $\mathbb{Q}$ representing u. By omitting finitely many terms of t, we can assume that each $|t_n|_p = 0$, that is, each t_n is a rational number a_n/b_n with $p \nmid a_n b_n$. This implies that b_n is a unit in the ring $\mathbb{Z}/p^n\mathbb{Z}$ of integers modulo p^n, so there exists $c_n \in \{1, \ldots, p^n - 1\}$, coprime to p, such that $b_n c_n = 1 - p^n e_n$ for some integer e_n. By construction, the terms of the sequence $z = (z_n)_{n\in\mathbb{N}}$ defined by $z_n = a_n c_n$ satisfy

$$|\tfrac{a_n}{b_n} - z_n|_p = |\tfrac{a_n - b_n a_n c_n}{b_n}|_p = |\tfrac{p^n a_n e_n}{b_n}|_p \leqslant p^{-n}.$$

It follows that z is a Cauchy sequence that defines the same element of $\mathbb{Q}_p$ as u, that is, $s = p^c z$. Since z contains an equivalent subsequence $(x_n)_{n\in\mathbb{N}}$ with $x_{n+1} \equiv x_n \bmod p^n$, we can assume that $z \in \mathbb{Z}_p$. □

Another useful representation of the elements of $\mathbb{Z}_p$ is to use sequences $(z_n)_{n\in\mathbb{N}}$ whose terms are the partial sums $z_n = a_0 + a_1 p + \ldots + a_n p^n$ of formal series $\sum_{i=0}^{\infty} a_i p^i$ where each $a_i \in \{0, \ldots, p-1\}$; clearly, $z_n \equiv z_{n+1} \bmod p^{n+1}$ for all n.

The next proposition shows that the elements with p-adic absolute value 1 are exactly the units in the ring $\mathbb{Z}_p$.

Proposition 3.7 *The* p-adic units *are* $\mathbb{Z}_p^* = \{s \in \mathbb{Q}_p \mid |s|_p = 1\}$.

Proof If $s \in \mathbb{Z}_p^*$, then both s and s^{-1} are elements of $\mathbb{Z}_p$ and have p-adic absolute value at most 1 by Proposition 3.6. Now $1 = |ss^{-1}|_p = |s|_p |s^{-1}|_p$ forces $|s|_p = 1$, so $\mathbb{Z}_p^*$ is a subset of $\{s \in \mathbb{Q}_p \mid |s|_p = 1\}$. Conversely, if $s \in \mathbb{Q}_p$ has p-adic absolute value 1, then $s \in \mathbb{Z}_p$ by Proposition 3.6, and so we can assume that s is represented by a sequence $s = (s_n)_{n\in\mathbb{N}}$ with $s_n \in \{0, \ldots, p^n - 1\}$ and $s_n = (s_{n+1} \bmod p^n)$ for all n. Since $|s|_p = 1$, each s_n is a unit modulo p^n, and there exists a unique $t_n \in \{1, \ldots, p^{n-1}\}$ with $s_n t_n \equiv 1 \bmod p^n$. By construction, $t = (t_n)_{n\in\mathbb{N}}$ satisfies $st = (1)_{n\in\mathbb{N}}$, and it remains to show that $t \in \mathbb{Z}_p$. This is obvious: if $n > 1$, then $s_n t_n \equiv 1 \bmod p^n$ implies $s_n t_n \equiv 1 \bmod p^{n-1}$; since $s_{n-1} = (s_n \bmod p^{n-1})$, we deduce that $s_{n-1} t_n \equiv 1 \bmod p^{n-1}$, so $t_n \bmod p^{n-1}$ is the unique inverse of s_{n-1} modulo p^{n-1}, that is, $t_{n-1} = (t_n \bmod p^{n-1})$. It follows that $t \in \mathbb{Z}_p$, and hence s is a unit in $\mathbb{Z}_p$. □

This and Lemma 3.4 imply that every nonzero $s \in \mathbb{Q}_p$ can be written uniquely as $s = p^c u$ for some $c \in \mathbb{Z}$ and unit $u \in \mathbb{Z}_p^*$.

We now consider the unit group $\mathbb{Z}_p^*$ in more detail. The first result is a consequence of Hensel's Lemma (see [65, Lemma II.4.6]) and shows that the p-adic integers contain a primitive $(p-1)$-th root of unity, that is, $\omega \in \mathbb{Z}_p$ of multiplicative order $p-1$, see [43, Chapter 10.3] for a proof.

Lemma 3.8 *Let $r \in \{2, \ldots, p-1\}$ be a primitive root modulo p. The element $\omega = \lim_{n\to\infty} r^{(p^n)}$ is a primitive $(p-1)$-th root of unity in $\mathbb{Z}_p$ with $\omega \equiv r^{(p^{n-1})} \bmod p^n$ for all $n \in \mathbb{N}$.*

The polynomial $\mathbf{x}^{p-1} - 1$ has $p-1$ distinct roots in in the field $\mathbb{Z}/p\mathbb{Z}$, and so it also has $p-1$ distinct roots in $\mathbb{Z}_p$ by Hensel's Lemma. In particular, the unit group of $\mathbb{Z}_p$ contains a cyclic subgroup μ_{p-1} of order $p-1$ consisting of all $(p-1)$-th roots of unity. If $p = 2$, then the only roots of unity in $\mathbb{Z}_2$ are ± 1; if $p > 2$, then the only roots of unity in $\mathbb{Z}_p$ are those in μ_{p-1}. We refer to [71, p. 51, Propositions 1 and 2] for a proof.

If follows from Proposition 3.7 that $u \in \mathbb{Z}_p$ is a unit if and only if $u \notin p\mathbb{Z}_p$. Thus, the reduction modulo p induces a surjective group homomorphism from $\mathbb{Z}_p^*$ onto $(\mathbb{Z}/p\mathbb{Z})^*$ whose kernel is the multiplicative group of *one-units* $1 + p\mathbb{Z}_p$. Together with Lemma 3.8, this shows that

$$\mathbb{Z}_p^* = \langle \omega \rangle \times (1 + p\mathbb{Z}_p)$$

where $\omega \in \mathbb{Z}_p^*$ is a primitive $(p-1)$-th root of unity; note that if $p = 2$, then $\mathbb{Z}_2^* = 1 + 2\mathbb{Z}_p$. To understand the multiplicative structure of $1 + p\mathbb{Z}_p$, one can use the formal p-adic exponential and logarithm maps, defined as

$$\exp(s) = 1 + s + \frac{s^2}{2!} + \frac{s^3}{3!} + \ldots \quad \text{and} \quad \log(1+s) = x - \frac{s^2}{2} + \frac{s^3}{3} - \frac{s^4}{4} + \ldots.$$

The proof of the following result can be found in [42, Proposition 4.5.9].

Proposition 3.9 *If $p > 2$, then* log *defines an isomorphism $(1+p\mathbb{Z}_p, \cdot) \longrightarrow (p\mathbb{Z}_p, +)$; if $p = 2$, then* log *defines an isomorphism $(1+4\mathbb{Z}_2, \cdot) \longrightarrow (4\mathbb{Z}_2, +)$. In both cases, the inverse is defined by* exp.

Next, we summarise some algebraic properties of $\mathbb{Z}_p$. Let I be a nonzero ideal in $\mathbb{Z}_p$ and choose a nonzero $s \in I$ such that $c = v_p(s)$ is as small as possible. By Lemma 3.4 and Proposition 3.7, we can write $s = p^c u$ with $u \in \mathbb{Z}_p^*$, hence $p^c = su^{-1} \in I$. The minimality of c implies that every other element of I is a multiple of p^c. This implies the following.

Lemma 3.10 *The ring $\mathbb{Z}_p$ is a principal ideal domain whose nonzero ideals are exactly the terms of the series*

$$\mathbb{Z}_p \rhd p\mathbb{Z}_p \rhd p^2\mathbb{Z}_p \rhd p^3\mathbb{Z}_p \rhd \ldots.$$

Each quotient $(p^i\mathbb{Z}_p/p^{i+1}\mathbb{Z}_p, +)$ is cyclic of order p, and $\mathbb{Z}_p/p\mathbb{Z}_p$ is isomorphic to the field $\mathbb{Z}/p\mathbb{Z}$ with p elements. In particular, the field of fractions of the integral domain $\mathbb{Z}_p$ is isomorphic to the field $\mathbb{Q}_p$.

The relevance of the p-adic integers in the study of pro-p-groups is underpinned by the observation that every abelian pro-p-group is a $\mathbb{Z}_p$-module.

Lemma 3.11 *An abelian pro-p-group G is a $\mathbb{Z}_p$-module, where a p-adic integer $z = (z_n)_{n\in\mathbb{N}}$ acts on $g \in G$ via $g^z = \lim_{n\to\infty} g^{z_n}$.*

Proof Let G be the inverse limit of the inverse system $\{G_i \mid i \in \Lambda\}$ of finite p-groups. Let $g = (g_i)_{i\in\Lambda} \in G$ and let $z = (z_n)_{n\in\mathbb{N}} \in \mathbb{Z}_p$. Since $z_n \equiv z_m \bmod p^n$ for all $m \geqslant n$, we deduce that $g_i^{z_n} = g_i^{z_m}$ whenever the order of G_i divides p^n. This implies that $\lim_{n\to\infty} g^{z_n}$ exists in G. A similar consideration shows that if $g, h \in G$ and $y, z \in \mathbb{Z}_p$, then $g^{y+z} = g^y g^z$ and $(gh)^z = g^z h^z$. It follows that $g \mapsto g^z$ is a well-defined group action of $\mathbb{Z}_p$ on G. □

3.3 The field $\mathbb{Q}_p$ and its finite extensions

We include a brief introduction to finite field extensions of $\mathbb{Q}_p$. For details and proofs we refer to [65, Chapters II.3–II.5]. We first note that the p-adic absolute value $|\cdot|_p$ of $\mathbb{Q}_p$ extends to certain finite extensions, see [65, Theorem II.4.8].

Proposition 3.12 *Let K be a finite extension of $\mathbb{Q}_p$ of degree $d = [K : \mathbb{Q}_p]$ and let λ_k be the $\mathbb{Q}_p$-linear map $K \to K$, $h \mapsto kh$. For $k \in K$ define*

$$|k|_p = (|\det \lambda_k|_p)^{1/d}.$$

This is the unique absolute value on K that extends the absolute value of $\mathbb{Q}_p$; it is non-archimedean, multiplicative, and makes K a complete field.

Similar to the previous section, we now define the *valuation ring*

$$\mathcal{O} = \{s \in K \mid |s|_p \leqslant 1\}.$$

Its unit group is

$$\mathcal{O}^* = \{s \in K \mid |s|_p = 1\},$$

and it has a unique maximal ideal

$$\mathfrak{p} = \{s \in K \mid |s|_p < 1\}.$$

The ring $\mathcal{O}$ is an integral domain with field of fractions (isomorphic to) K. The field $\mathcal{O}/\mathfrak{p}$ is the *residue class field* of $\mathcal{O}$. Moreover, $\mathcal{O}$ is a principal ideal domain and there exists a prime element κ of $\mathcal{O}$ such that the nonzero ideals of $\mathcal{O}$ are exactly

$$\mathcal{O} \rhd \mathfrak{p} \rhd \mathfrak{p}^2 \rhd \mathfrak{p}^3 \rhd \dots \quad \text{where} \quad \mathfrak{p}^z = \kappa^z\mathcal{O} \text{ for } z \in \mathbb{Z}, \tag{3.1}$$

and every quotient $\mathfrak{p}^z/\mathfrak{p}^{z+1}$ is isomorphic to $\mathcal{O}/\mathfrak{p}$. Every finite extension of $\mathbb{Q}_p$, such as K above, is a *local field*; these are fields that are complete with respect to a discrete absolute value and have a finite residue class field. For these fields the following information about their multiplicative structure is available, see [65, Satz II.3.10, II.5.3 and II.5.5]. For an integer $n \geqslant 1$, the group of n-th *one-units* is

$$\mathcal{U}^{(n)} = 1 + \mathfrak{p}^n.$$

Proposition 3.13 *Let K, $\mathcal{O}$, κ, and $\mathfrak{p}$ be as above. Let q be the number of elements of the residue class field $\mathcal{O}/\mathfrak{p}$, and let $e \geqslant 1$ be defined by $p\mathcal{O} = \mathfrak{p}^e$.*

a) *There are decompositions $K^* = \langle\kappa\rangle \times \mathcal{O}^*$ and $\mathcal{O}^* = C_{q-1} \times \mathcal{U}^{(1)}$.*

b) *If $n > e/(p-1)$, then there is an isomorphism $(\mathcal{U}^{(n)}, \cdot) \to (\mathfrak{p}^n, +)$ induced by* log*; its inverse is induced by* exp.

c) *If $n \geqslant 1$, then $\mathcal{O}^*/\mathcal{U}^{(n)} \cong (\mathcal{O}/\mathfrak{p}^n)^*$ and $\mathcal{U}^{(n)}/\mathcal{U}^{(n+1)} \cong (\mathcal{O}/\mathfrak{p}, +)$.*

Note that if $K = \mathbb{Q}_p$, then these statements cover the results presented for $\mathbb{Q}_p$ and $\mathbb{Z}_p$ in Section 3.2.

3.4 Cyclotomic extensions of $\mathbb{Q}_p$

We now consider special cases of field extensions, specifically, cyclotomic extensions. We will focus on cyclotomic extensions adjoining a p-th root of unity.

3.4.1 General cyclotomic extensions

In the special case that K is a cyclotomic extension of $\mathbb{Q}_p$, the following statement [65, Satz II.7.13] holds:

Proposition 3.14 *Let θ be a primitive p^r-th root over $\mathbb{Q}_p$, and set $d = (p-1)p^{r-1}$.*

a) *The extension $\mathbb{Q}_p(\theta)/\mathbb{Q}_p$ has degree d and cyclic Galois group of size d.*

b) *The valuation ring of $\mathbb{Q}_p(\theta)$ is $\mathcal{O} = \mathbb{Z}_p[\theta]$, the smallest subring of $\mathbb{Q}_p$ that contains $\mathbb{Z}_p$ and θ, and $\kappa = \theta - 1$ is a prime element of $\mathbb{Z}_p[\theta]$.*

If θ is a primitive p^r-th root of unity over $\mathbb{Q}_p$, then Proposition 3.14 shows that $\{1, \theta, \ldots, \theta^{d-1}\}$ is a $\mathbb{Q}_p$-basis of $\mathbb{Q}_p(\theta)$ where $d = (p-1)p^{r-1}$. In particular, θ is a root of the (irreducible) polynomial

$$\frac{\mathbf{x}^{p^r} - 1}{\mathbf{x}^{p^{r-1}} - 1} = 1 + \mathbf{x}^{p^{r-1}} + \mathbf{x}^{2p^{r-1}} + \ldots + \mathbf{x}^{(p-2)p^{r-1}} + \mathbf{x}^d, \tag{3.2}$$

that is, $1 + \theta^{p^{r-1}} + \theta^{2p^{r-1}} + \ldots + \theta^{(p-2)p^{r-1}} + \theta^d = 0$. It follows that

$$\mathbb{Z}_p[\theta] = \{a_0 + a_1\theta + \ldots + a_{d-1}\theta^{d-1} \mid a_0, \ldots, a_{d-1} \in \mathbb{Z}_p\}$$

is a free $\mathbb{Z}_p$-module of rank d. It is a principal ideal domain whose nonzero ideals are exactly

$$\mathbb{Z}_p[\theta] \rhd \mathfrak{p} \rhd \mathfrak{p}^2 \rhd \ldots \quad \text{where} \quad \mathfrak{p}^z = \kappa^z\mathcal{O} \text{ for } \kappa = \theta - 1.$$

Since each $\mathfrak{p}^z/\mathfrak{p}^{z+1}$ is isomorphic to $\mathbb{Z}_p[\theta]/\mathfrak{p} \cong \mathbb{Z}/p\mathbb{Z}$, it follows that $p\mathfrak{p}^k = \mathfrak{p}^{k+d}$ for every k. In particular, $p\mathbb{Z}_p[\theta] = \mathfrak{p}^d$, and Proposition 3.13 implies the following.

Proposition 3.15 *If θ is a primitive p^r-th root of unity over $\mathbb{Q}_p$ and $\omega \in \mathbb{Z}_p$ is a primitive $(p-1)$-th root of unity, then*

$$\mathbb{Z}_p[\theta]^* = \langle \omega \rangle \times \mathcal{U}^{(1)}$$

and $\mathcal{U}^{(1)}/\mathcal{U}^{(2)}$ is cyclic of order p, generated by $\theta\mathcal{U}^{(2)}$. If $n > p^{r-1}$, then the exponential and logarithm maps define mutually inverse isomorphisms between the multiplicative group $\mathcal{U}^{(n)}$ and the additive group $(\mathfrak{p}^n, +)$.

We now give a short proof for the claims on the group $S(p^r)$ considered in Example 3.1.

Proposition 3.16 *Let θ be a primitive p^r-th root of unity over $\mathbb{Q}_p$. Let $P = \langle \theta \rangle$ be the cyclic group of order p^r that acts via multiplication on $T = (\mathbb{Z}_p[\theta], +)$. The associated semidirect product is isomorphic to the group $S(p^r)$ defined in Example 3.1, that is,*

$$S(p^r) \cong P \ltimes T;$$

in particular, $S(p^r)$ is an infinite pro-p-group of coclass r.

Proof Write $S = S(p^r)$. By Proposition 3.14 and the subsequent discussion, T has $\mathbb{Z}_p$-basis $B = \{1, \theta, \ldots, \theta^{d-1}\}$ where $d = (p-1)p^{r-1}$. Representing the elements of T by their coefficient vectors with respect to B shows that $T \cong (\mathbb{Z}_p^d, +)$. Multiplication by θ defines a $\mathbb{Z}_p$-linear map on T whose matrix representation with respect to B is the companion matrix c of the cyclotomic polynomial (3.2); in particular, $P \cong \langle c \rangle$. It follows that $P \ltimes T$ is isomorphic to the split extension $\langle c \rangle \ltimes (\mathbb{Z}_p^d, +)$, which is precisely the group S. This proves the first claim.

Write $T_j = \gamma_{j+1}(S)$ for $j \in \mathbb{N}$. The group $\gamma_2(S)$ is generated by all commutators

$$[(\theta^i, s), (\theta^j, t)] = [1, s(\theta^j - 1) - t(\theta^i - 1)]$$

where $s, t \in T$ and $i, j \in \{0, \ldots, p^r - 1\}$. Recall that $\mathfrak{p} = (\theta - 1)T$ is the principal the ideal of T generated by $\theta - 1$, and $(\theta^n - 1) = (\theta - 1)(1 + \theta + \ldots + \theta^{n-1})$ for $n \geqslant 1$. Hence, each of the above commutators lies in $\{1\} \times \mathfrak{p}$. On the other hand, if $s \in T$, then $[(1, s), (\theta, 0)] = (1, s(\theta - 1))$, which allows us to deduce that $\gamma_2(S) = \{1\} \times \mathfrak{p}$. An induction on j now shows that $\gamma_j(S) = \{1\} \times \mathfrak{p}^{j-1}$ for all $j \geqslant 2$, that is, $\gamma_j(S) \cong (\mathfrak{p}^{j-1}, +)$. In particular, $[\gamma_j(S) : \gamma_{j+1}(S)] = p$ for every $j \geqslant 2$, and therefore S is an infinite pro-p-group of coclass r, as claimed. □

3.4.2 Adjoining a p-th root of unity

We discuss other number theoretic results that will be important later when studying groups of coclass 1. Throughout, let p be an odd prime and consider $K = \mathbb{Q}_p(\theta)$ where θ is a primitive p-th root of unity. (Note that if $p = 2$, then $\theta = -1$ and $\mathbb{Q}_2(\theta) = \mathbb{Q}_2$.) We write $\mathcal{O} = \mathbb{Z}_p[\theta]$ with unique maximal ideal $\mathfrak{p} = \kappa\mathcal{O}$, where

$\kappa = \theta - 1$. We also write $d = p - 1$. Recall that $p^y\mathfrak{p}^z = \mathfrak{p}^{z+yd}$ for all $y, z \in \mathbb{Z}$, the extension $K/\mathbb{Q}_p$ has degree d, and $\mathcal{O}$ is a free $\mathbb{Z}_p$-module of dimension d.

We write $\mathcal{U} = \mathbb{Z}_p[\theta]^*$ for the unit group, and $\mathcal{U}^{(i)} = 1 + \mathfrak{p}^i$ for the i-th group of one-units. We have seen above that every nonzero $s \in K$ can be written uniquely as $s = \kappa^z u$ for some $z \in \mathbb{Z}$ and $u \in \mathcal{U}$; we use this to define $\mathrm{val}(s)$ by the property $s \in \mathfrak{p}^{\mathrm{val}(s)} \setminus \mathfrak{p}^{\mathrm{val}(s)+1}$. This yields a map

$$\mathrm{val}\colon K^\times \to \mathbb{Z},\ s \mapsto \mathrm{val}(s).$$

Note that if $s, t \in K$ are nonzero, then $\mathrm{val}(st) = \mathrm{val}(s) + \mathrm{val}(t)$.

Proposition 3.14 shows that the Galois group $\mathrm{Gal}(K/\mathbb{Q}_p)$ is cyclic of order d, comprising elements $\sigma_1, \ldots, \sigma_d$, each defined by $\theta^{\sigma_i} = \theta^i$. We fix a generator, say

$$\mathrm{Gal}(K/\mathbb{Q}_p) = \langle\sigma\rangle \quad \text{with} \quad \theta^\sigma = \theta^r,$$

where $r \in \{1, \ldots, p-1\}$ is a primitive root modulo p. By Lemma 3.8, we can assume that our fixed primitive d-th root of unity $\omega \in \mathbb{Z}_p$ satisfies $\omega \equiv r \bmod p$.

Lemma 3.17 *With the previous notation, the following hold.*

a) *As a $\mathbb{Q}_p$-linear map, the Galois automorphism σ is diagonalisable with eigenvalues $\omega^0, \ldots, \omega^{d-1}$, and each eigenspace has dimension 1.*

b) *The eigenvalue of an eigenvector $u \in K$ of σ is $\omega^{\mathrm{val}(u)} = \omega^{\mathrm{val}(u) \bmod d}$.*

c) *For every $z \in \mathbb{Z}$ there exists an eigenvector u of σ with* $\mathrm{val}(u) = z$.

d) *If k is a positive divisor of d, then σ^k has d/k distinct eigenspaces of dimension k each, with eigenvalues $\{\omega^{ki} \mid i \in \{0, \ldots, d-1\}\}$.*

Proof a) Since σ has order d, the minimal polynomial $\mu_\sigma(\mathrm{x})$ of σ over $\mathbb{Q}_p$ divides $\mathrm{x}^d - 1 = (\mathrm{x}-\omega^0)(\mathrm{x}-\omega^1)\cdots(\mathrm{x}-\omega^d)$, where $(\mathrm{x}^d-1)/(\mathrm{x}-1) = 1+\mathrm{x}+\ldots+\mathrm{x}^{d-1}$ is irreducible over $\mathbb{Q}_p$ by Proposition 3.14. Since $1 \in K$ is an eigenvector of σ with eigenvalue 1, we deduce that $\mu_\sigma(\mathrm{x}) = \mathrm{x}^d - 1$. The $\mathbb{Q}_p$-dimension of K is d, which implies the claim.

b) If $\mathrm{val}(u) = k$, then $u = c\kappa^k + \kappa^{k+1}u'$ for some $c \in \{1, \ldots, p-1\}$ and $u' \in \mathcal{O}$. Replacing u by $c^{-1}u$, we can assume that $u \equiv \kappa^k \bmod \mathfrak{p}^{k+1}$. Recall that $\theta^\sigma = \theta^r$ and $\theta \equiv 1 \bmod \mathfrak{p}$, and we assume $\omega \equiv r \bmod p$. It follows that

$$u^\sigma \equiv (\theta^r - 1)^k \equiv (1+\theta+\ldots+\theta^{r-1})^k(\theta-1)^k \equiv r^k\kappa^k \equiv \omega^k u \bmod \mathfrak{p}^{k+1}.$$

This implies that the eigenvalue of u must be ω^k, as claimed.

c) Part a) shows that there exist eigenvectors $u_0, u_1, \ldots, u_d$ of σ such that each u_i has corresponding eigenvalue ω^i. Part b) shows that each $\mathrm{val}(u_i) = k_i$ satisfies $i \equiv k_i \bmod d$. If $z \in \mathbb{Z}$ is given, then $z \equiv k_i \bmod d$ for a unique i, and we can write $z = k_i + yd$ for some $y \in \mathbb{Z}$. Now $u = p^y u_i$ is an eigenvector of σ with eigenvalue ω^i and $\mathrm{val}(u) = k_i + yd = z$.

d) This follows from part a). □

Lemma 3.17 allows us to make the following definition: for every $k \in \mathbb{N}$ choose

$$v_k \in \mathfrak{p}^k \setminus \mathfrak{p}^{k+1} \quad \text{such that} \quad \sigma(v_k) = \omega^k v_k.$$

Proposition 3.18 *For every integer $z \geqslant 0$, the set $\{v_z, v_{z+1}, \dots, v_{z+d-1}\}$ is $\mathbb{Q}_p$-linearly independent and a $\mathbb{Z}_p$-basis of $\mathfrak{p}^z$.*

Proof Since $v_z, \dots, v_{z+d-1}$ have distinct eigenvalues, a standard linear algebra argument show that they are $\mathbb{Q}_p$-linearly independent, so they form a basis of K. Since each $v_i \in \mathfrak{p}^i \setminus \mathfrak{p}^{i+1}$ and $\mathfrak{p}^{i+d} = p\mathfrak{p}^i$ for every i, the second claim follows. □

The next result was proved by Leedham-Green & McKay [56]; we use the convention that the subscript of σ_a is read modulo p.

Proposition 3.19 *For $a \in \{2, \dots, d\}$ define*

$$\tau_a \colon \mathcal{O} \to \mathcal{O}, \quad c \mapsto c^{\sigma_a} + c^{\sigma_{1-a}} - c.$$

Considered as a $\mathbb{Q}_p$-linear transformation of K, the following hold.

a) *If $a \equiv r^i \bmod p$ and $1-a \equiv r^j \bmod p$, then the eigenvalues of τ_a are $\omega^{ik}+\omega^{jk}-1$ for $k \in \{0, \dots, d-1\}$.*

b) *If $p \equiv 5 \bmod 6$, then $\ker \tau_a$ is trivial for every $a \in \{2, \dots, d\}$.*

c) *If $p \equiv 1 \bmod 6$, then there exists $a \in \{2, \dots, d\}$ such that $\ker \tau_a$ has dimension $d/3$.*

Proof a) Since $\theta^\sigma = \theta^r$, the assumptions imply that $\sigma_a = \sigma^i$ and $\sigma_{1-a} = \sigma^j$. If v_k is one of the basis vectors of K defined above, then $v_k^\sigma = \omega^k v_k$, and therefore

$$v_k^{\tau_a} = \omega^{ki} v_k + \omega^{kj} v_k - v_k = (\omega^{ik} + \omega^{jk} - 1) v_k.$$

This shows that each v_k is an eigenvector of τ_a with eigenvalue $\omega^{ik} + \omega^{jk} - 1$. Since $v_0, \dots, v_{d-1}$ form a $\mathbb{Q}_p$-basis of K, the claim follows.

b) If the eigenvalues of τ_a are as in a), then the dimension of $\ker \tau_a$ is the number of $k \in \{0, \dots, d-1\}$ with $\omega^{ik} + \omega^{jk} = 1$. For such a value of k, we use Proposition 3.5 and consider ω^{ik}, 0, and 1 as complex numbers. Since each of ω^{ik}, ω^{jk}, and 1 has complex norm 1, and $\omega^{ik} + \omega^{ik} = 1$ by assumption, the elements 0, 1, and ω^{ik} are the vertices of an equilateral triangle in the complex plane. This is possible if and only if the angle in the complex plane between 1 and ω^{ik} is $60°$, that is, if and only if ω^{ik} is a primitive 6-th root of unity. This implies that $6 \mid d$ and $p \equiv 1 \bmod 6$. Thus, if $p \equiv 5 \bmod 6$, then this case is not possible, and so τ_a has trivial kernel.

c) The assumptions imply that there exists a primitive 6-th root modulo p, say $a \in \{2, \dots, d\}$. Recall the assumption that $r \equiv \omega \bmod p$, so if $i \in \{0, \dots, d\}$ satisfies $a \equiv r^i \bmod p$, then ω^i is a primitive 6-th root of unity. It follows from Proposition 3.5 that in this case $\omega^i + \omega^{-i} = 1$. Thus, $1 - a \equiv 1 - \omega^i \equiv \omega^{-i} \bmod p$, and part a) shows that τ_a has eigenvalues $\omega^{ik} + \omega^{-ik} - 1$ for $k \in \{0, \dots, d-1\}$. The proof of part b) shows that $\omega^{ik} + \omega^{-ik} = 1$ if and only if ω^{ik} is a primitive 6-th root of unity. Since ω^i has order 6, the latter holds if and only if $\gcd(k, 6) = 1$. In $\{0, 1, \dots, d-1\}$ there are $d/2$ even integers and $d/6$ odd integers divisible by 3. Thus, τ_a has $d - d/2 - d/6 = d/3$ eigenvalues 0. □

If τ_a is as in Proposition 3.19, then we denote its eigenvalues by

$$\omega_{a,k} = \omega^{ik} + \omega^{jk} - 1 \quad (k \in \{0, \dots, d-1\}).$$

For $k \in \{2, \dots, p\}$ let V_k be the intersection of $\mathfrak{p}^2$ with the eigenspace of σ with eigenvalue ω^k, that is, V_k is the $\mathbb{Z}_p$-span of the vector v_k. It follows that

$$\mathfrak{p}^2 = V_2 \oplus \dots \oplus V_p$$

is a direct sum of cyclic $\mathbb{Z}_p$-modules, and τ_a acts on each V_k by multiplication by the eigenvalue $\omega_{a,k}$. We now translate this structure to the multiplicative group $\mathcal{U}^{(2)}$ with the help of the isomorphism $\exp\colon (\mathfrak{p}^2, +) \to \mathcal{U}^{(2)}$ of Proposition 3.15. This yields a direct product decomposition

$$\mathcal{U}^{(2)} = A_2 \times \dots \times A_p$$

where each $A_k = \exp(V_k) = \{\exp(v_k)^z \mid z \in \mathbb{Z}_p\}$. Since σ is a field automorphism, it is compatible with the exponential and logarithm maps. This allows us to translate the action of τ_a on $\mathfrak{p}^2$ to an action of $\mathcal{U}^{(2)}$: for $a \in \{2, \dots, d\}$ let

$$\rho_a\colon \mathcal{U} \to \mathcal{U}, \quad u \mapsto u^{\sigma_a} u^{\sigma_{1-a}} u^{-1}. \tag{3.3}$$

By construction, $\exp(s)^{\rho_a} = \exp(s^{\tau_a})$ for every $s \in \mathfrak{p}^2$. This shows that each subgroup A_k is invariant under ρ_a, and ρ_a acts on A_k by raising elements to the power $\omega_{a,k}$.

We conclude this section by recording some properties of a group homomorphism $\mathcal{U} \to \mathcal{U}$ that will be relevant later.

Proposition 3.20 *Let $k > 0$ be a divisor of d and set $\tau = \sigma^k$, which has order $h = d/k$.*

a) *The homomorphism*

$$\chi\colon \mathcal{U} \to \mathcal{U}, \quad u \mapsto u(u^{-1})^{\tau},$$

satisfies $\mathcal{U} = \ker(\chi) \times \mathrm{im}(\chi)$.

b) *If $h = d$, then* $\ker(\chi) = \mathbb{Z}_p^*$. *If $h < d$, then*

$$\ker(\chi) = \langle \omega \rangle \times (1 + \mathcal{F})$$

where $\mathcal{F} = \{a \in \mathfrak{p}^2 \mid a^{\tau} = a\}$ is a $\mathbb{Z}_p$-subspace of dimension k in $\mathcal{O}$.

Proof a) It is clear that χ is a group homomorphism. Proposition 3.15 shows that $\mathcal{U} = \langle \omega \rangle \times \langle \theta \rangle \times \mathcal{U}^{(2)}$. Note that $\omega^{\chi} = 1$ and $\theta^{\chi} = \theta^{1-r^k}$. We now consider the restriction of χ to $\mathcal{U}_2 = 1 + \mathfrak{p}^2$. The exponential map translates χ to a map $\psi\colon \mathfrak{p}^2 \to \mathfrak{p}^2$, $x \mapsto x - x^{\tau}$. It follows from Lemma 3.17 that ψ is diagonalisable with h distinct eigenspaces, each of $\mathbb{Z}_p$-dimension k. In particular, this shows that $\mathfrak{p}^2 = \ker(\psi) \oplus \mathrm{im}(\psi)$, and therefore $\mathcal{U}_2$ is the direct product of the kernel and image of the restriction $\chi|_{\mathcal{U}_2}$.

b) This follows from Lemma 3.17; note that if $h = d$, then $\tau = \sigma$ and the fixed points of τ in K are exactly $\mathbb{Q}_p$; hence $\ker(\chi) = \mathbb{Q}_p \cap \mathcal{U} = \mathbb{Z}_p^*$. □

Chapter 4
Matrix groups and uniseriality

In Chapter 1, we established that the maximal coclass trees in $\mathcal{G}(p,r)$ correspond one-to-one to the isomorphism types of infinite pro-p-groups of coclass r, see Proposition 1.24. The aim of this chapter is to initiate the investigation of the structure of these infinite pro-p-groups.

We begin with two central ingredients: the notion of uniseriality and an analysis of the finite p-groups acting uniserially on $\mathbb{Z}_p^d$. This setup is used to construct and investigate p-adic space groups and, based on this, infinite pro-p-groups of finite coclass. Many results in this chapter are known, but in some cases difficult to find in the literature. We include a detailed treatment for the convenience of the reader.

4.1 Uniseriality and two-step centralisers

If a finite p-group G acts on a finite abelian p-group M, then there exists a series of G-invariant normal subgroups $M = M_1 > M_2 > \ldots > M_\ell > M_{\ell+1} = \{1\}$ such that G centralises each section M_i/M_{i+1} of this series, see Proposition 1.4. In the following we investigate the extremal case when there is exactly one such centralised series. For a group G acting on a finite abelian group M, we write $[G,M]$ for the subgroup of M defined as

$$[G,M] = \langle m^g m^{-1} \mid m \in M, g \in G \rangle.$$

If $g \in G$ and $m \in M$, then we also write $[g,m] = (m^{-1})^g m$; this generalises the usual notation of commutators in groups.

Definition 4.1 Let G be a finite p-group acting on a finite abelian p-group M. Then G acts *uniserially* on M if $[G,N]$ has index p in N for every non-trivial G-invariant subgroup N of M.

The next proposition explains the terminology 'uniserial': there exists a unique series of G-invariant subgroups.

H. Dietrich and B. Eick, *Coclass graphs of p-groups*, Springer Monographs in Mathematics, https://doi.org/10.1007/978-3-032-17506-9_4

Proposition 4.2 *Let G be a finite p-group acting on a finite abelian p-group M of order p^k. Define $M_0 = M$ and $M_{i+1} = [G, M_i]$ for $i \in \{0, \dots, k-1\}$. The following are equivalent:*

a) *The group G acts uniserially on M.*

b) *The series $M = M_0 > M_1 > \dots$ satisfies $[M_i : M_{i+1}] = p$ for $i \in \{0, \dots, k-1\}$.*

c) *The subgroups $M_0, M_1, \dots, M_k$ are the only G-invariant subgroups of M.*

Proof We prove this by a chain of implications.

a) $\Rightarrow$ b): An induction on i shows that each M_i is G-invariant. Now $[M_i : M_{i+1}] = p$ for $0 \leqslant i \leqslant k-1$ follows from the definition of uniseriality.

b) $\Rightarrow$ c): Suppose that there is a G-invariant subgroup $N \leqslant M$ outside the series. Then there exists an index i such that $N \nleqslant M_{i+1}$ and $N \leqslant M_i$; since N is non-trivial and M_{k-1} has order p, it follows that $i \leqslant k-2$. The subgroup $K = NM_{i+2}$ of M is G-invariant with $M_{i+2} \leqslant K \leqslant M_i$ and $K \nleqslant M_{i+1}$. Since M_i/M_{i+2} has order p^2, this implies that M_i/M_{i+2} is the direct product of K/M_{i+2} and M_{i+1}/M_{i+2}. Both subgroups are G-invariant of order p, hence G acts trivially on M_i/M_{i+2}. But then $[G, M_i] \leqslant M_{i+2}$, a contradiction. Hence, such an N cannot exist.

c) $\Rightarrow$ a): By construction, $M_0 \geqslant \dots \geqslant M_k$. This implies that $M_0 = M$ and $M_k = \{1\}$. Since M has order p^k, it follows that $M_{i+1} = [G, M_i]$ has index p in M_i for every $i \in \{0, \dots, k-1\}$, proving uniseriality. □

Next, let G be a group that acts on an infinite abelian pro-p-group M. Recall from Lemma 3.11 that M is a $\mathbb{Z}_pG$-module, where $\mathbb{Z}_pG$ denotes the group ring of G with coefficients in $\mathbb{Z}_p$. We denote by $[G, M]$ the subgroup of M that is the $\mathbb{Z}_pG$-submodule of M generated by all commutators $[g, m] = (m^{-1})^g m$; we also write

$$[G, M] = \langle m^g m^{-1} \mid m \in M, g \in G \rangle_{\mathbb{Z}_p}.$$

Motivated by Proposition 4.2b), we define the following.

Definition 4.3 Let G be a group acting on an infinite abelian pro-p-group M, and define $M_0 = M$ and $M_{i+1} = [G, M_i]$ for $i \in \mathbb{N}_0$. The G-action on M is *uniserial* if $[M_i : M_{i+1}] = p$ for all $i \in \mathbb{N}_0$ and the intersection of all $M_0, M_1, \dots$ is the trivial subgroup. In this case, $M_0 > M_1 > \dots$ is the *associated uniserial series*.

An example of a uniserial action on an infinite pro-p-group is exhibited by the p-adic group $S(p^r) = P \ltimes T$ in Section 3.1: the proof of Proposition 3.16 reveals that the finite group P of order p^r acts uniserially on the infinite pro-p-group $T = (\mathbb{Z}_p^d, +)$ where $d = (p-1)p^{r-1}$.

If a group G acts uniserially on an infinite abelian pro-p-group M with uniserial series $M = M_0 > M_1 > \dots$, then G acts uniserially on each finite abelian quotient M/M_i; in this sense, the definitions of uniserial action in the finite and the infinite case are compatible. The next result is the analogue of Proposition 4.2 for uniserial actions on infinite pro-p-groups.

Proposition 4.4 *Let G be a group acting uniserially on an infinite abelian pro-p-group M, with uniserial series $M = M_0 > M_1 > \ldots$.*

a) *The subgroups $M_0, M_1, \ldots$ are the only non-trivial $\mathbb{Z}_p G$-submodules of M.*

b) *Any two non-trivial $\mathbb{Z}_p G$-submodules of M have a non-trivial intersection.*

c) *If $M \cong (\mathbb{Z}_p^d, +)$ for some $d \in \mathbb{N}$, then $M_i^{[p]} = M_{i+d}$ for every $i \in \mathbb{N}_0$.*

Proof a) Suppose N is a $\mathbb{Z}_p G$-invariant subgroup of M outside the series. Then there exists i such that $N \not\leqslant M_{i+1}$ and $N \leqslant M_i$. As in the proof of Proposition 4.2b), the subgroup $K = NM_{i+2}$ is $\mathbb{Z}_p G$-invariant with $M_{i+2} \leqslant K \leqslant M_i$ and $K \not\leqslant M_{i+1}$. The same argument in that proof establishes the contradiction $[G, M_i] \leqslant M_{i+2}$. Thus, such an N does not exist.

b) If $A, B \leqslant M$ are non-trivial $\mathbb{Z}_p G$-submodules, then $A = M_i$ and $B = M_j$ for some $i, j \in \mathbb{N}_0$ by a), and therefore $A \cap B = M_{\max(i,j)}$ is non-trivial.

c) The group $M_i^{[p]}$ is $\mathbb{Z}_p G$-invariant and has index p^d in M_i. The same holds for M_{i+d}, hence $M_i^{[p]} = M_{i+d}$ by a). □

Hand-in-hand with uniseriality goes the concept of a two-step centraliser.

Definition 4.5 Let G be a group acting uniserially on a (finite or infinite) abelian pro-p-group M with series $M = M_0 > M_1 > \ldots$. If $M_{i+1} \neq \{1\}$, then the i-th *two-step centraliser* of G on M is $C_G(M_i/M_{i+2}) = \{g \in G \mid [g, M_i] \subseteq M_{i+2}\}$.

Let G be a group that acts uniserially on an abelian pro-p-group M. It follows from the definition that a subgroup of a two-step centraliser does not act uniserially on M. Moreover, the 'converse' is true: if a subgroup of G is not contained in any two-step centraliser, then it acts uniserially. The next proposition provides more details on two-step centralisers.

Proposition 4.6 *Let $R \leqslant P \leqslant G$ be finite p-groups such that G acts on an abelian pro-p-group M (finite or infinite). If the induced action of P on M is uniserial, then the following hold.*

a) *Each two-step centraliser of P on M is a maximal subgroup of P.*

b) *The group R acts uniserially on M if and only if R is not contained in any two-step centraliser of P on M.*

c) *The group G acts uniserially on M.*

Proof Let $M = M_0 > M_1 > \ldots$ denote the uniserial series of P on M.

a) Let $i \in \mathbb{N}_0$ with $M_{i+1} \neq \{1\}$ and consider $Q_i = M_i/M_{i+2}$. This group has order p^2, and therefore is cyclic or elementary abelian. In both cases, $\mathrm{Aut}(Q_i)$ has a cyclic Sylow p-subgroup of order p. Since P is uniserial, it acts non-trivially on Q_i. Thus the image of the induced action homomorphism $P \longrightarrow \mathrm{Aut}(Q_i)$ has order p and its kernel has index p in P. This kernel is precisely the two-step centraliser $C_P(Q_i)$ and hence $C_P(Q_i)$ is a maximal subgroup.

b) Since each $[R, M_i] \leqslant [P, M_i]$, it follows that R acts uniserially on M if and only if $[R, M_i] = [P, M_i]$ for every $i \in \mathbb{N}_0$. Clearly, if R is contained in some two-step centraliser, then it does not act uniserially. For the converse, note that each M_i is $\mathbb{Z}_p R$-invariant. If R is not contained in any two-step centraliser, then the same argument as in the proofs of Propositions 4.2 and 4.4 establishes that $M_0, M_1, \ldots$ are *all* the $\mathbb{Z}_p R$-invariant submodules of M, in particular, each $[R, M_i] = [P, M_i]$.

c) Recall that P is subnormal in G (see Proposition 1.5), that is, there exist subgroups $P = P_0 \leqslant P_1 \leqslant \ldots \leqslant P_k = G$ such that each P_i is normal in P_{i+1}. If we establish that P_1 acts uniserially on M, then an inductive argument proves the claim for G. Thus, it is sufficient to assume that P is normal in G. In this case, another induction proves that each $M_{j+1} = [P, M_j]$ is G-invariant, and so there is an induced G-action on M_j/M_{j+1}. If the latter is non-trivial, then it has order p, and so G acts trivially on M_j/M_{j+1} by Proposition 1.4, that is, $[G, M_j] \leqslant M_{j+1}$. Since $M_{j+1} = [P, M_j]$ by assumption, it follows that $[G, M_j] = M_{j+1}$. This implies the claim. □

We conclude this section with another useful observation.

Lemma 4.7 *Let G be a group acting uniserially on a (finite or infinite) abelian pro-p-group M with series $M = M_0 > M_1 > \ldots$. Let $u \in M_i \setminus M_{i+1}$ for some $i \in \mathbb{N}_0$. Then u generates M_i as $\mathbb{Z}_p G$-module.*

Proof Let V be the smallest $\mathbb{Z}_p G$-submodule of M containing the element u. Propositions 4.2 and 4.4 show that $V = M_j$ for some j. Since $u \in M_i \setminus M_{i+1}$, it follows that $i = j$. □

4.2 Construction of finite p-groups with uniserial action

A prototype example of a uniserial action is a finite p-subgroup P of $\mathrm{GL}_d(\mathbb{Z}_p)$ that acts uniserially on the natural $\mathbb{Z}_p$-module, that is, on the infinite abelian pro-p-group $T = (\mathbb{Z}_p^d, +)$. In the following we call such a subgroup P a *uniserial p-subgroup* of $\mathrm{GL}_d(\mathbb{Z}_p)$. Our aim is to determine the maximal uniserial p-subgroups in $\mathrm{GL}_d(\mathbb{Z}_p)$ up to conjugacy in $\mathrm{GL}_d(\mathbb{Q}_p)$ and in $\mathrm{GL}_d(\mathbb{Z}_p)$, respectively.

4.2.1 Conjugacy classes in $\mathrm{GL}_d(\mathbb{Q}_p)$

We start with a well-known construction for some important examples of finite uniserial p-subgroups of $\mathrm{GL}_d(\mathbb{Z}_p)$; see [57, Chapter 10] for more details.

Proposition 4.8 *Let $P \leqslant \mathrm{GL}_d(\mathbb{Z}_p)$ be a finite p-group acting uniserially on $(\mathbb{Z}_p^d, +)$. Then $(\mathbb{Q}_p^d, +)$ is an irreducible (or simple) $\mathbb{Q}_p P$-module.*

Proof Write $T = (\mathbb{Z}_p^d, +)$ and $V = (\mathbb{Q}_p^d, +)$. If W is a non-trivial $\mathbb{Q}_p P$-submodule of V, then $0 < \dim(W) \leqslant d$ and the intersection $L = T \cap W$ is a non-trivial $\mathbb{Z}_p P$-submodule of T. Proposition 4.4 now shows that L is part of the uniserial series of P on T, in particular, L has finite index in T. This forces $\dim(W) = d$, hence $W = V$, as claimed. □

A central ingredient of our desired construction is the *regular wreath product* $A \wr B$ of a group A and a finite group B. To define this, we note that the group B acts on itself by right multiplication and this induces a homomorphism φ from B into the symmetric group Sym_n for $n = |B|$. We define $A \wr B = B \ltimes A^n$ where $v \in B$ acts on A^n by permuting the n direct factors A according to $\varphi(v)$, that is, if $(u; r_1, \ldots, r_n), (v; s_1, \ldots, s_n) \in A \wr B$, then

$$(u; r_1, \ldots, r_n)(v; s_1, \ldots, s_n) = (uv; r_{1\varphi(v)^{-1}} s_1, \ldots, r_{n\varphi(v)^{-1}} s_n).$$

We use the convention $A \wr B \wr C = (A \wr B) \wr C$ for iterated wreath products.

If P is a subgroup of $\mathrm{GL}_d(\mathbb{Z}_p)$ and C_n is a cyclic group of order n, then the wreath product $P \wr C_n$ can be considered as a subgroup of $\mathrm{GL}_{nd}(\mathbb{Z}_p)$: the elements of the base group $P^n = P \times \ldots \times P$ are block-diagonal matrixes of dimension $nd \times nd$, whose $d \times d$ blocks are the elements of P, and a generator of C_n can be defined by the Kronecker product of the $n \times n$ permutation matrix representing an n-cycle and the $d \times d$ identity matrix. This construction preserves uniseriality, as the next proposition shows.

Proposition 4.9 *If $P \leqslant \mathrm{GL}_d(\mathbb{Z}_p)$ is a finite p-group acting uniserially on $(\mathbb{Z}_p^d, +)$, then $P \wr C_p \leqslant \mathrm{GL}_{dp}(\mathbb{Z}_p)$ is a finite p-group acting uniserially on $(\mathbb{Z}_p^{pd}, +)$.*

Proof Write $T = (\mathbb{Z}_p^d, +)$ and $V = (\mathbb{Z}_p^{pd}, +)$. Let g be a generator of C_p and let $G = P \wr C_p \leqslant \mathrm{GL}_{dp}(\mathbb{Z}_p)$. The corresponding series of P and G are constructed as usual, that is, $T_0 = T$ and $V_0 = V$, and $T_{i+1} = [P, T_i]$ and $V_{i+1} = [G, V_i]$ for $i \in \mathbb{N}_0$. By hypothesis, $[T_i : T_{i+1}] = p$ for each i and $\bigcap_{i \in \mathbb{N}_0} T_i = \{0\}$. We show that the same holds for the terms V_i.

We first describe the basic idea of the proof. Let $B \cong P \times \ldots \times P$ denote the base group of the wreath product G. Then B acts diagonally on V and thus $[B, V]$ is a subgroup of index p^p in V. The cyclic group C_p acts on the quotient $V/[B, V]$ as a p-th root of unity. This induces a uniserial action on this quotient and thus yields a series $V = V_0 > V_1 > \ldots > V_p = [B, V]$. This series is the initial part of the series through V and it satisfies $[V_i : V_{i+1}] = p$ for $i \in \{0, \ldots, p-1\}$. Iterating this process with $V_p, V_{2p}, \ldots$ yields the desired result.

We now provide details of the proof. Decompose $V = T_{1,0} \oplus \ldots \oplus T_{p,0}$, where each $T_{j,0}$ is a copy of T with uniserial series $T_{j,0} > T_{j,1} > \ldots$, such that g permutes the subspaces $T_{1,0}, \ldots, T_{p,0}$ cyclically. We write elements of V as $v = (v_1, \ldots, v_p)$ with each $v_j \in T_{j,0}$. Similarly, denote by $P_j \leqslant P^p$ the direct factor that acts on $T_{j,0}$, and write $h \in P^p$ as $h = (h_1, \ldots, h_p)$ with each $h_j \in P_j$. For $v \in V$ and $k \in \mathbb{N}$ denote $[g(k), v] = [g, [g(k-1), v]]$ with $[g(1), v] = [g, v]$. If $v = (v_1, 0, \ldots, 0) \in V$ and $h = (h_1, 1, \ldots, 1) \in P^p$, then

$$[h,[g(1),v]] = [h,(v_1,-v_1,0,\ldots,0)] = ([h_1,v_1],0,\ldots,0),$$
$$[h,[g(2),v]] = [h,(v_1,-2v_1,v_1,0,\ldots,0)] = ([h_1,v_1],0,\ldots,0),$$

and, more generally, $[h,[g(k),v]] = ([h_1,v_1],0,\ldots,0)$ for each $k \in \{1,\ldots,p-1\}$. By construction, the elements $[h,[g(p-1),v]]$ lie in V_p and, moreover, they generate $T_{1,1}$. The analogous argument shows that $T_{j,1} \leqslant V_p$ for each $j = 1,\ldots,p$. Since $T_{1,1} \oplus \ldots \oplus T_{p,1} \leqslant V_p$ has index p^p in V, we deduce that $[V : V_p] \leqslant p^p$. Repeating this argument, an induction on i shows that $[V : V_{ip}] \leqslant p^{ip}$ for every $i \in \mathbb{N}$. We also know that $V_j > V_{j+1}$ for every $j \in \mathbb{N}$, which follows from the observation that for each $k \in \mathbb{N}$ the group $(P \wr C_p) \ltimes (V/p^k V)$ is a finite p-group, hence nilpotent. Together, this implies that $[V_j : V_{j+1}] = p$ for every $j \in \mathbb{N}$. The construction also shows that $\bigcap_{i\in\mathbb{N}_0} V_i = \{0\}$, that is, $P \wr C_p$ acts uniserially on V, as claimed. □

Proposition 4.9 allows us to construct new uniserial p-subgroups of $\mathrm{GL}_{pd}(\mathbb{Z}_p)$ from those in $\mathrm{GL}_d(\mathbb{Z}_p)$, and this construction can be iterated. The base case for the inductive procedure is the family of primitive groups: these are matrix groups whose natural module cannot be decomposed as a proper direct sum of subspaces that are permuted by the action of the group. We refer to [57, Theorem 10.1.25] for a proof of the following proposition.

Proposition 4.10 *Let $G \leqslant \mathrm{GL}_d(\mathbb{Z}_p)$ be a finite p-group that acts primitively on $(\mathbb{Q}_p^d,+)$. Then either G is cyclic of order p and $d = p-1$, or G is isomorphic to the quaternion group Q_{16} and $d = 4$.*

The cyclic p-subgroups of $\mathrm{GL}_d(\mathbb{Z}_p)$ can be investigated in a little more generality. Recall from Proposition 3.14 (and the discussion thereafter) that if θ is a primitive p^s-th root of unity over $\mathbb{Q}_p$, then the degree of the field extension $\mathbb{Q}_p(\theta)/\mathbb{Q}_p$ is $d = (p-1)p^{s-1}$, and the cyclic group $P = \langle\theta\rangle$ acts uniserially on $T = (\mathbb{Z}_p[\theta],+)$. As $(\mathbb{Z}_p[\theta],+)$ is isomorphic to $(\mathbb{Z}_p^d,+)$, this shows that P can be identified with a subgroup of $\mathrm{GL}_d(\mathbb{Z}_p)$ that acts uniserially on $(\mathbb{Z}_p^d,+)$. By Proposition 4.8, this group is irreducible. The next proposition shows that these are essentially all the cyclic irreducible subgroups of $\mathrm{GL}_d(\mathbb{Q}_p)$ of p-power order.

Proposition 4.11 *If $G \leqslant \mathrm{GL}_d(\mathbb{Q}_p)$ is an irreducible cyclic subgroup of order p^s, then $d = (p-1)p^{s-1}$ and G is conjugate to the subgroup $C_{(p,s)}$ generated by the companion matrix of the p^s-th cyclotomic polynomial. The group $C_{(p,s)}$ is a subgroup of $\mathrm{GL}_d(\mathbb{Z}_p)$ and acts uniserially on $(\mathbb{Z}_p^d,+)$.*

Proof Consider the irreducible representations of the cyclic group of order p^s. These are, up to equivalence, in one-to-one correspondence with the conjugacy classes of subgroups of $\mathrm{GL}_d(\mathbb{Q}_p)$ generated by matrices of order p^s that do not preserve any proper non-trivial subspace. Let M be such a matrix. Its minimal polynomial is the irreducible cyclotomic polynomial $\mu(\mathbf{x}) = 1 + \mathbf{x}^{p^{s-1}} + \ldots + \mathbf{x}^{(p-1)p^{s-1}}$ described in Equation (3.2). Now linear algebra implies that the minimal dimension of a irreducible subspace is at least $(p-1)p^{s-1}$. On the other hand, if v is a nonzero vector of the natural $\mathrm{GL}_d(\mathbb{Q}_p)$-module, then the cyclic M-invariant subspace generated by v (that is, spanned by $\{v, vM, vM^2,\ldots\}$) has dimension at most $(p-1)p^{s-1}$.

Since $\langle M \rangle$ is irreducible, this forces $d = (p-1)p^{s-1}$ and $\langle M \rangle$ is conjugate to the subgroup $C_{(p,s)} \leqslant \mathrm{GL}_d(\mathbb{Z}_p)$ generated by the companion matrix of $\mu(\mathbf{x})$. The explicit construction via a primitive p^s-th root of unity θ (see the discussion prior to the proposition) proves that the action is uniserial. □

Our next construction requires some preliminary observations. First, the polynomial $\mathbf{x}^2 + \mathbf{x} + 4$ has a root in $\mathbb{Z}_2/2\mathbb{Z}_2$. By Hensel's Lemma (see [65, Lemma II.4.6]), it follows that $\mathbf{x}^2 + \mathbf{x} + 4$ has a root v in $\mathbb{Z}_2$, and now $u = 1 + 2v \in \mathbb{Z}_2^*$ satisfies $u^2 = -15$. We denote u by $\sqrt{-15}$. This allows the following construction which makes the non-cyclic case in Proposition 4.10 explicit.

Proposition 4.12 *If $G \leqslant \mathrm{GL}_4(\mathbb{Q}_2)$ is a primitive subgroup isomorphic to Q_{16}, then G is conjugate to $\langle a, b \rangle$ where*

$$a = \frac{1}{\sqrt{-15}} \begin{pmatrix} 2 & 1 & 1 & 3 \\ 1 & 1 & 3 & -2 \\ 1 & 3 & -2 & -1 \\ 3 & -2 & -1 & -1 \end{pmatrix} \quad \text{and} \quad b = \begin{pmatrix} 0 & 1 & 0 & 0 \\ 0 & 0 & 1 & 0 \\ 0 & 0 & 0 & 1 \\ -1 & 0 & 0 & 0 \end{pmatrix}.$$

The group $\langle a, b \rangle$ is a subgroup of $\mathrm{GL}_4(\mathbb{Z}_2)$ that acts uniserially on $(\mathbb{Z}_2^4, +)$.

Proof The proof requires some 2-adic representation theory, which we only cite. First, it follows from [57, Lemma 10.1.19] that $\langle a, b \rangle$ is a faithful irreducible representation of Q_{16}. Second, the uniqueness up to conjugacy follows from [57, Theorem 10.1.22]. Lastly, the proof of [57, Theorem 10.1.25] shows that a faithful irreducible representation of Q_{16} over $\mathbb{Q}_2$ is primitive. □

Next, we use these primitive groups in a wreath product construction.

Definition 4.13 Fix a prime p, and let $s \in \mathbb{N}$ and $t \in \mathbb{N}_0$.

a) We write $d_s = (p-1)p^{s-1}$, and assume throughout that the prime p is known from the context.

b) Let $W_1(t) = C_{(p,t)}$ as defined in Proposition 4.11, and for $s > 1$ let $W_s(t) = W_{s-1}(t) \wr C_p$. Then $W_s(t)$ is a subgroup of $\mathrm{GL}_{d_{t+s-1}}(\mathbb{Z}_p)$. We also write $W_s = W_s(1) \leqslant \mathrm{GL}_{d_s}(\mathbb{Z}_p)$.

c) We write $V_3 = Q_{16} \leqslant \mathrm{GL}_{d_3}(\mathbb{Z}_2)$ for the group defined in Proposition 4.12, and define $V_s = V_{s-1} \wr C_2 \leqslant \mathrm{GL}_{d_s}(\mathbb{Z}_2)$ for $s \geqslant 4$.

Propositions 4.9, 4.11, and 4.12 show that the groups in Definition 4.13 act uniserially on their natural modules. The groups W_s and V_s play a particularly important role as demonstrated by the next proposition.

Note that the set of subgroups of $\mathrm{GL}_d(\mathbb{Z}_p)$ is not closed under conjugation by $\mathrm{GL}_d(\mathbb{Q}_p)$. We define the $\mathbb{Q}_p$-*class* of a subgroup P in $\mathrm{GL}_d(\mathbb{Z}_p)$ as the intersection of $\mathrm{GL}_d(\mathbb{Z}_p)$ with the $\mathrm{GL}_d(\mathbb{Q}_p)$-conjugacy class of P, that is, the $\mathbb{Q}_p$-class of P consists of all conjugates P^g with $g \in \mathrm{GL}_d(\mathbb{Q}_p)$ such that $P^g \in \mathrm{GL}_d(\mathbb{Z}_p)$.

Proposition 4.14 *If* $\mathrm{GL}_d(\mathbb{Z}_p)$ *has an irreducible subgroup of p-power order, then* $d = d_s$ *for some* $s \in \mathbb{N}$. *Moreover, the following hold.*

a) *For odd p, or if* $s < 3$, *there is a unique* $\mathbb{Q}_p$*-class in* $\mathrm{GL}_{d_s}(\mathbb{Z}_p)$ *of irreducible p-subgroups of largest possible p-power order, with representative* W_s.

b) *For* $p = 2$ *and* $s \geqslant 3$ *there are two* $\mathbb{Q}_2$*-classes in* $\mathrm{GL}_{d_s}(\mathbb{Z}_2)$ *of irreducible subgroups of largest possible* 2*-power order, with representatives* W_s *and* V_s.

Proof We prove a) and b) simultaneously. Write $T = (\mathbb{Z}_p^d, +)$ and $V = (\mathbb{Q}_p^d, +)$, and let $P \leqslant \mathrm{GL}_d(\mathbb{Z}_p)$ be an irreducible subgroup of largest possible p-power order. First suppose that P acts imprimitively on V, that is, P preserves a direct sum decomposition $V = V_1 \oplus \ldots \oplus V_n$ of $n > 1$ subspaces of dimension d/n which are permuted by P. Since P is irreducible, P permutes these subspaces transitively. Let B be the image of P under the associated permutation representation $P \to \mathrm{Sym}_n$. We realise Sym_n as a subgroup of $\mathrm{GL}_d(\mathbb{Z}_p)$ consisting of permutation matrices that permute the subspaces $V_1, \ldots, V_n$. The group B is a finite p-group, and transitivity implies that $n = p^e$ for some e (by the Orbit-Stabiliser Theorem). If S is the stabiliser of V_1 in P, considered as a subgroup of $\mathrm{GL}(V_1) = \mathrm{GL}_{d/n}(\mathbb{Q}_p)$, then P is a subgroup of the wreath product $S \wr B$. The maximality of P means that $P = S \wr B$ where $B \leqslant \mathrm{Sym}_{p^e}$ is a Sylow p-subgroup and $S \leqslant \mathrm{GL}_{d/n}(\mathbb{Z}_p)$ is an irreducible subgroup that is maximal with respect to being a finite p-group. As B is a Sylow subgroup of a symmetric group, it follows that B is isomorphic to an iterated wreath product $C_p \wr \ldots \wr C_p$ with e copies of C_p, see for example [72, (1.6.19)]. Repeating this consideration for S, a finite iteration shows that $P = K \wr C_p \wr C_p \wr \ldots \wr C_p$ with $t > 0$ copies of C_p, and $K \leqslant \mathrm{GL}_m(\mathbb{Z}_p)$ a finite, irreducible, primitive p-subgroup. The claims now follow from Proposition 4.10 and Definition 4.13. □

This settles the classification, up to $\mathrm{GL}_d(\mathbb{Q}_p)$-conjugacy, of the irreducible finite subgroups of $\mathrm{GL}_d(\mathbb{Z}_p)$ of largest possible p-power order. Since every finite uniserial p-subgroup of $\mathrm{GL}_d(\mathbb{Z}_p)$ is irreducible, we observe the following.

Corollary 4.15 *The groups listed in Proposition 4.14 form a complete and irredundant set of maximal uniserial p-subgroups of* $\mathrm{GL}_d(\mathbb{Z}_p)$ *up to* $\mathrm{GL}_d(\mathbb{Q}_p)$*-conjugacy.*

4.2.2 Conjugacy classes in $\mathrm{GL}_d(\mathbb{Z}_p)$

We have seen in the previous section that, up to $\mathrm{GL}_d(\mathbb{Q}_p)$-conjugacy, every finite uniserial p-subgroup of $\mathrm{GL}_d(\mathbb{Z}_p)$ is a subgroup of W_s or of V_s for some s with $d = d_s$. Our next aim is to describe how the $\mathbb{Q}_p$-classes split into different $\mathrm{GL}_d(\mathbb{Z}_p)$-classes.

Our main tool is Proposition 4.16. This result seems folklore; we include a proof for completeness. For example, this result has also been used in the classification of ordinary space groups, see Opgenorth, Plesken & Schulz [69].

We say a subgroup $L \leqslant (\mathbb{Q}_p^d, +)$ is a *lattice* if L is a free $\mathbb{Z}_p$-module that contains a $\mathbb{Q}_p$-basis of $\mathbb{Q}_p^d$; this is sometimes called a full lattice. The group $\mathrm{GL}_d(\mathbb{Q}_p)$ acts on the set of all lattices by right multiplication.

Proposition 4.16 *Let $P \leqslant \mathrm{GL}_d(\mathbb{Z}_p)$ be a finite p-subgroup acting uniserially on $T = (\mathbb{Z}_p^d, +)$ with series $T = T_0 > T_1 > \ldots$. Let $\mathcal{L} = \{p^i T_j \mid i \in \mathbb{Z}, 0 \leqslant j < d\}$ and write $N = N_{\mathrm{GL}_d(\mathbb{Q}_p)}(P)$ for the normaliser of P in $\mathrm{GL}_d(\mathbb{Q}_p)$.*

a) *The elements of $\mathcal{L}$ are exactly the P-invariant lattices in $\mathbb{Q}_p^d$, and N acts on $\mathcal{L}$ by right multiplication.*

b) *There is a bijection between the N-orbits in $\mathcal{L}$ and the $\mathrm{GL}_d(\mathbb{Z}_p)$-conjugacy classes contained in the $\mathbb{Q}_p$-class of P.*

Proof a) Proposition 4.4 implies that $\{T_i \mid i \in \mathbb{N}_0\} = \{p^i T_j \mid i \in \mathbb{N}_0, 0 \leqslant j < d\}$ is the set of P-invariant lattices in $\mathbb{Z}_p^d$. If L is a P-invariant lattice in $\mathbb{Q}_p^d$, then $p^k L \in \{T_i \mid i \in \mathbb{N}_0\}$ for some $k \in \mathbb{N}$. This implies the first claim. For the second claim, consider $L \in \mathcal{L}$ and $x \in N$. Since x normalises P, it follows that

$$(Lx)g = L(xg) = L(g^{x^{-1}}x) = (Lg^{x^{-1}})x \subseteq Lx$$

for each $g \in P$. This shows that $Lx \in \mathcal{L}$, and it follows that N acts on $\mathcal{L}$.

b) Let $g_1, \ldots, g_s \in \mathrm{GL}_d(\mathbb{Q}_p)$ such that $P^{g_1^{-1}}, \ldots, P^{g_s^{-1}}$ is a complete and irredundant list of $\mathrm{GL}_d(\mathbb{Z}_p)$-class representatives in the $\mathbb{Q}_p$-class of P. We assume $g_1 = 1$.

Let $T = \mathbb{Z}_p^d$ and consider $\mathcal{R} = \{Tg_1, \ldots, Tg_s\}$. Since each $P^{g_i^{-1}} \leqslant \mathrm{GL}_d(\mathbb{Z}_p)$, if $g \in P$, then $h = g_i g g_i^{-1}$ lies in $\mathrm{GL}_d(\mathbb{Z}_p)$, and therefore $Tg_i g = Thg_i \subseteq Tg_i$. This implies $Tg_i \in \mathcal{L}$, hence $\mathcal{R} \subseteq \mathcal{L}$. Next, we show that distinct elements of $\mathcal{R}$ lie in distinct N-orbits. For this purpose, suppose that $Tg_i x = Tg_j$ for some $x \in N$. Then $g_i x g_j^{-1}$ stabilises T, and so $g_i x = hg_j$ for some $h \in \mathrm{GL}_d(\mathbb{Z}_p)$. Since $x \in N$, this shows that $P^{g_j^{-1}h^{-1}} = P^{x^{-1}g_i^{-1}} = P^{g_i^{-1}}$, so $P^{g_i^{-1}}$ and $P^{g_j^{-1}}$ are conjugate under $\mathrm{GL}_d(\mathbb{Z}_p)$. By the definition of $g_1, \ldots, g_s$, this forces $g_i = g_j$.

Finally, we show that each $L \in \mathcal{L}$ is contained in an N-orbit of some element of $\mathcal{R}$. Note that we can write $L = Th$ for some $h \in \mathrm{GL}_d(\mathbb{Q}_p)$. Since L is P-invariant, every $g \in P$ satisfies $(Th)g = Lg = L = Th$, so $hgh^{-1} \in \mathrm{GL}_d(\mathbb{Z}_p)$. This shows that $hPh^{-1} \leqslant \mathrm{GL}_d(\mathbb{Z}_p)$ lies in the $\mathbb{Q}_p$-class of P, and therefore $P^{h^{-1}}$ is $\mathrm{GL}_d(\mathbb{Z}_p)$-conjugate to $P^{g_i^{-1}}$ for a unique $i \in \{1, \ldots, s\}$, say $P^{h^{-1}} = P^{g_i^{-1}k}$ for some $k \in \mathrm{GL}_d(\mathbb{Z}_p)$. This implies that $x = g_i^{-1}kh$ is in N, and now $L = Th = Tk^{-1}g_i x = Tg_i x$ proves that L lies in the N-orbit of an element of $\mathcal{R}$.

This proves the claim; the bijection maps the $\mathrm{GL}_d(\mathbb{Z}_p)$-class with representative $P^{g_i^{-1}}$ to the N-orbit in $\mathcal{L}$ with representative Tg_i. □

Thus, to construct all uniserial p-subgroups in $\mathrm{GL}_d(\mathbb{Z}_p)$ with $d = d_s$ up to conjugacy, we proceed as follows. Up to $\mathrm{GL}_d(\mathbb{Q}_p)$-conjugacy, every such subgroup is contained in W_s or in V_s, so the first step is to determine $\mathrm{GL}_d(\mathbb{Q}_p)$-class representatives $P_1, \ldots, P_k \leqslant \mathrm{GL}_d(\mathbb{Z}_p)$ of the uniserial subgroups in W_s and V_s. The second step is to determine, for each $P = P_i$, the distinct $\mathrm{GL}_d(\mathbb{Z}_p)$-class representatives in the $\mathbb{Q}_p$-class of P. Proposition 4.16 shows that this can be solved by determining the $N_{\mathrm{GL}_d(\mathbb{Q}_p)}(P)$-orbits in the set of P-invariant lattices. It remains to investigate the action of this normaliser. As before, we denote by $T = T_0 > T_1 > \ldots$ the uniserial series associated to the uniserial action of P. In the following, we write

$$N_{\mathbb{Q}_p}(P) = N_{\mathrm{GL}_d(\mathbb{Q}_p)}(P),$$
$$C_{\mathbb{Q}_p}(P) = C_{\mathrm{GL}_d(\mathbb{Q}_p)}(P),$$
$$\mathcal{L}(P) = \{p^i T_j \mid i \in \mathbb{Z}, 0 \leqslant j < d\}.$$

It is shown in Proposition 4.16 that $\mathcal{L}(P)$ is the set of all P-invariant lattices in $\mathbb{Q}_p^d$. The next observation asserts that a set of $N_{\mathbb{Q}_p}(P)$-orbit representatives in $\mathcal{L}(P)$ is finite: this is due to the fact that multiplication by p can be realised by a scalar (hence central) matrix. We summarise this in the following lemma.

Lemma 4.17 *Let $P \leqslant \mathrm{GL}_d(\mathbb{Z}_p)$ be a finite uniserial p-subgroup acting uniserially on $T = (\mathbb{Z}_p^d, +)$ with series $T = T_0 > T_1 > \ldots$. The set $\{T_0, \ldots, T_{d-1}\}$ contains a representative for each $N_{\mathbb{Q}_p}(P)$-orbit in $\mathcal{L}(P)$.*

Proposition 4.18 *Let $P \leqslant \mathrm{GL}_d(\mathbb{Z}_p)$ be a finite uniserial p-subgroup with $d = d_s = (p-1)p^{s-1}$. Write $T = (\mathbb{Z}_p^d, +)$ with uniserial series $T = T_0 > T_1 > \ldots$.*

a) *The centre $Z(P)$ of P is cyclic of order dividing p^s.*

b) *If $g \in C_{\mathbb{Q}_p}(P)$ has order p^j, then $j \leqslant s$ and g is $\mathrm{GL}_d(\mathbb{Q}_p)$-conjugate to a block diagonal matrix with p^{s-j} diagonal blocks, and each block acts as a primitive p^j-th root of unity via the matrix group $C_{(p,j)}$ as described in Proposition 4.11.*

c) *Let I be the identity in $\mathrm{GL}_d(\mathbb{Q}_p)$. If there exists $g \in C_{\mathbb{Q}_p}(P)$ of order p^j, then $x = g - I$ lies in $C_{\mathbb{Q}_p}(P)$ and satisfies $T_i x = T_{i+p^{s-j}}$ for all $i \in \mathbb{N}_0$.*

Proof Write $V = (\mathbb{Q}_p^d, +)$ for the natural $\mathbb{Q}_p P$-module.

a) Recall that P acts irreducibly on V by Proposition 4.8. Now Schur's Lemma [61, Lemma 1.3.2] shows that the endomorphism ring $E = \mathrm{End}_P(V)$ of P-module homomorphisms $V \to V$ is a division algebra. Multiplication by $z \in Z(P)$ defines a P-module homomorphism in E and, since P acts faithfully on V, we can identify $Z(P)$ with an abelian subgroup of the multiplicative group of E. If F denotes the division subalgebra of E generated by the centre of E and the subgroup $Z(P)$ of E, then F is abelian, hence a field. As every finite subgroup of the multiplicative group of a field is cyclic (see for example [46, Proposition 2.80]), it follows that $Z(P)$ is cyclic. The claim on the order follows from b).

b) Let $\tilde{P}$ be the group generated by P and g. Since $g \in C_{\mathbb{Q}_p}(P)$ has finite order, it follows that $\tilde{P}$ is finite and $g \in Z(\tilde{P})$. Note that V is an irreducible $\mathbb{Q}_p\tilde{P}$-module since P acts irreducibly. If Z denotes the (normal) subgroup of $\tilde{P}$ generated by g, then Clifford's Theorem [47, Theorem 6.5] shows that $V = W_1 \oplus \ldots \oplus W_\ell$ is a direct sum of irreducible $\mathbb{Q}_p Z$-modules such that $\dim(W_1) = \ldots = \dim(W_\ell)$. Since Z acts as an irreducible cyclic group on each W_i, Proposition 4.11 shows that this action is, up to conjugacy, via $C_{(p,j)}$. It also implies that $\dim(W_i) = (p-1)p^{j-1}$, and so $\ell = p^{s-j}$. This forces $j \leqslant s$, which completes the proof.

c) Part b) shows that, up to conjugacy, g is a block diagonal matrix of p^{s-j} blocks, each equal to the companion matrix h of a primitive p^j-th root of unity. We now show that $x = g - I$ has the claimed property. The characteristic polynomial of h is the p^j-th cyclotomic polynomial, see Equation (3.2). Evaluating this polynomial at

1 yields $\det(h - 1) = p$, which shows that $\det(x) = p^{(p^{s-j})}$. Thus, x is invertible, and so $x \in C_{\mathbb{Q}_p}(P)$. Let D be the Smith normal form of x and write $x = LDR$ where $L, R \in \mathrm{GL}_d(\mathbb{Z}_p)$. We can assume that the diagonal entries of D are powers of p and thus $\det(D) = \det(x) = p^{(p^{s-j})}$. Note that $T_i g$ and $T_i x$ are P-invariant, and as L and R are invertible over $\mathbb{Z}_p$, it follows that $[T : T_i x] = [T : T_i D] = [T : T_i] \det(x)$. Thus,

$$[T : T_i x] = [T : T_i] p^{(p^{s-j})} = [T : T_{i+p^{s-j}}],$$

and Proposition 4.4a) proves that $T_i x = T_{i+p^{s-j}}$. □

We conclude this section with two results for the centraliser action.

Proposition 4.19 *Let $p > 2$, write $d = d_s$, and let $P \leqslant W_s$ a uniserial p-subgroup of $\mathrm{GL}_d(\mathbb{Z}_p)$ with associated uniserial series $T = T_0 > T_1 > \ldots$.*

a) *The $N_{\mathbb{Q}_p}(P)$-orbits on $\mathcal{L}(P)$ coincide with the $C_{\mathbb{Q}_p}(P)$-orbits on $\mathcal{L}(P)$.*

b) *If $|Z(P)| = p^c$, then $T_0, \ldots, T_{p^{s-c}-1}$ is a complete list of $N_{\mathbb{Q}}(P)$-orbit representatives.*

Proof a) It follows from [24, Theorem 25] and its proof that $N_{\mathbb{Q}_p}(P) = CN$ where the subgroup C (called $C_{i,\mathbb{Q}_p}(s)$ in [24]) centralises P and the subgroup N lies in a subgroup (called $N_i(s)$ in [24]) of $N_{\mathbb{Z}_p}(P)$. Thus, if $h \in N$, then every $T_i h$ is P-invariant and $[T : T_i h] = [T : T_i]$, so $T_i h = T_i$ by Proposition 4.4. This shows that $N \leqslant \mathrm{GL}_d(\mathbb{Z}_p)$ acts trivially on $\mathcal{L}(P)$, and so the $N_{\mathbb{Q}_p}(P)$-orbits coincide with the C-orbits. Since $C \leqslant C_{\mathbb{Q}_p}(P)$, the claim follows.

b) By Lemma 4.17 and part a), it remains to reduce $\{T_0, \ldots, T_d\}$ modulo the action of $C_{\mathbb{Q}_p}(P)$. If $|Z(P)| = p^c$, then Proposition 4.18a,c) proves the claim. □

Remark 4.20 In all known applications, the list of lattices in Proposition 4.19b) is irredundant. However, proving this seems a delicate matter.

Proposition 4.21 *Let $p = 2$ and write $d = d_s = 2^{s-1}$. Let $P \leqslant \mathrm{GL}_d(\mathbb{Z}_2)$ be a finite 2-subgroup that acts uniserially on $T = (\mathbb{Z}_2^d, +)$ with series $T = T_0 > T_1 > \ldots$.*

a) *If $s \geqslant 3$ and $P \leqslant V_s$, there exists $x \in C_{\mathbb{Q}_2}(P)$ with $T_i x = T_{i+2^{s-2}}$ for all $i \in \mathbb{N}_0$.*

b) *If $s \geqslant 2$ and $P = W_s$, there exists $x \in N_{\mathbb{Q}_2}(P)$ with $T_i x = T_{i+2^{s-2}}$ for all $i \in \mathbb{N}_0$.*

Proof a) We use induction on s. For $s = 3$, the group V_3 is described in Proposition 4.12, with generators a and b. Choose $x = b + b^{-1}$. Then $x^2 = 2I$, where I is the identity in V_3. A direct calculation shows that $xa = ax$ and $xb = bx$, so x centralises V_3, and therefore also P. Now observe that $T_{i+4} = 2T_i = T_i x^2 < T_i x < T_i$ for each $i \in \mathbb{N}_0$, which implies that $T_i x = T_{i+2}$, as claimed. Now let $s > 3$ and recall that $V_s = C_2 \ltimes (V_{s-1} \times V_{s-1})$. Let $x' \in C_{\mathbb{Q}_2}(V_{s-1})$ exist by the induction hypothesis, and define x as the block diagonal matrix with two identical blocks, x' and x'. By construction, $x \in C_{\mathbb{Q}_2}(V_s)$ has the required properties.

b) Again, we use induction on s. For $s = 2$, we consider

$$a = \begin{pmatrix} 0 & 1 \\ 1 & 0 \end{pmatrix}, \quad b = \begin{pmatrix} -1 & 0 \\ 0 & 1 \end{pmatrix}, \quad x = \begin{pmatrix} -1 & -1 \\ -1 & 1 \end{pmatrix},$$

and note that $W_2 = \langle a, b \rangle \cong D_8$ and $x \in N_{\mathbb{Q}_2}(W_2) \setminus C_{\mathbb{Q}_2}(W_2)$. Hence $x^2 = 2I$ where I is the 2×2 identity matrix, and so $T_{i+2} = 2T_i = T_i x^2 < T_i x < T_i$ for all $i \in \mathbb{N}_0$. It follows that $T_i x = T_{i+1}$ for all $i \in \mathbb{Z}$. The induction step is now analogous to a). □

4.3 Structure of finite p-groups with uniserial action

The previous section has established that if a finite p-subgroup P of $\mathrm{GL}_d(\mathbb{Z}_p)$ with uniserial action on $T = (\mathbb{Z}_p^d, +)$ exists, then $d = d_s = (p-1)p^{s-1}$ for some $s \in \mathbb{N}$, and P is conjugate in $\mathrm{GL}_d(\mathbb{Q}_p)$ to a subgroup of W_s or V_s. In Section 5.4, we exhibit bounds for s: for example, Theorem 5.32 shows how s is related to the coclass of P.

We now investigate the structure of W_s and V_s and their uniserial subgroups in more detail. This includes a detailed account of the two-step centralisers of W_s and V_s and a proof of the result that every uniserial subgroup of W_s or V_s can be generated by at most s or $s+1$ elements, respectively.

4.3.1 The uniserial subgroups of W_s

In Definition 4.13, we introduced W_s as a subgroup of $\mathrm{GL}_{d_s}(\mathbb{Z}_p)$. Here we briefly review this construction and provide some equivalent descriptions.

First, we define for $s \in \mathbb{N}$ the iterated regular wreath product $\hat{W}_s = C_p \wr \ldots \wr C_p$ of s copies of C_p. By construction, $\hat{W}_1 = C_p$ and

$$\hat{W}_s = \hat{W}_{s-1} \wr C_p = C_p \ltimes (\hat{W}_{s-1})^p,$$

where C_p acts regularly by permuting the p copies of $\hat{W}_{s-1}$. Induction shows that

$$|\hat{W}_s| = p^{1+p+p^2+\ldots+p^{s-1}}.$$

The abstract group $\hat{W}_s$ acts naturally on the set $\Omega_s = \{1, \ldots, p^s\}$. This action can be defined iteratively: if $\hat{W}_{s-1}$ acts on $\Omega_{s-1} = \{1, \ldots, p^{s-1}\}$, then the i-th copy of $\hat{W}_{s-1}$ in $\hat{W}_s = C_p \ltimes (\hat{W}_{s-1})^p$ acts on the shifted subset $(i-1)p^{s-1} + \Omega_{s-1}$ of Ω_s, and the subgroup C_p permutes the p copies of $\hat{W}_{s-1}$ and their action domains cyclically. This construction is well known and yields an embedding of $\hat{W}_s$ into Sym_{p^s} whose image is a Sylow p-subgroup of Sym_{p^s}. We denote this permutation representation by

$$\varphi_s \colon \hat{W}_s \longrightarrow \mathrm{Sym}_{p^s}.$$

Using permutation matrices, the permutation representation φ_s translates into a permutation matrix representation

$$\rho_s \colon \hat{W}_s \longrightarrow \mathrm{GL}_{p^s}(\mathbb{Z}_p).$$

We now introduce two representations of $\hat{W}_s$ that are induced by ρ_s. First, note that $\hat{W}_s = C_p \wr \hat{W}_{s-1} = \hat{W}_{s-1} \ltimes B$ with $B \cong C_p^{p^{s-1}}$ elementary abelian of rank p^{s-1}. The image $\rho_s(B)$ acts as block-diagonal matrix group with p^{s-1} independent diagonal blocks each of type C_p. Each block C_p acts on $\mathbb{Z}_p^p$ with 1-dimensional fixed-point space $\langle(1,\ldots,1)\rangle \leqslant \mathbb{Z}_p^p$, and the direct sum of all these spaces yields a fixed-point space F for $\rho_s(B)$ of dimension p^{s-1}. By construction, F is a $\hat{W}_s$-submodule, and the induced action of $\hat{W}_s$ on $\mathbb{Z}_p^{p^s}/F$ is denoted by

$$\sigma_s \colon \hat{W}_s \longrightarrow \mathrm{GL}_{d_s}(\mathbb{Z}_p).$$

Note that σ_s is injective, and its image is the matrix group W_s as constructed in Definition 4.13; this is the central object of our study here. Second, we use the natural homomorphism $\mathbb{Z}_p \longrightarrow \mathbb{Z}_p/p\mathbb{Z}_p \cong \mathbb{F}_p$, where $\mathbb{F}_p$ denotes the finite field with p elements. This translates ρ_s into the representation

$$\tau_s \colon \hat{W}_s \longrightarrow \mathrm{GL}_{p^s}(p).$$

In the following, we usually identify W_s with its abstract copy $\hat{W}_s$.

Recall that the rank of a finite p-group is the minimal generator number of the group; we determine the rank of W_s in the following lemma. For this denote by g_i a generator of the i-th factor C_p in the iterated wreath product W_s; we also denote this by writing

$$W_s = C_p(g_1) \wr \ldots \wr C_p(g_s).$$

Proposition 4.22 *For every $s \in \mathbb{N}$, the group W_s has rank s, and $\{g_1,\ldots,g_s\}$ is a minimal generating set of W_s.*

Proof Since $\{g_1,\ldots,g_s\}$ generates W_s, it follows that $\mathrm{rk}(W_s) \leqslant s$ for all s. We now use induction on s. Clearly, $\mathrm{rk}(W_1) = 1$, so we assume that $\mathrm{rk}(W_{s-1}) = s-1$ for some $s \geqslant 2$. As explained above, we can realise $W_s = W_{s-1} \ltimes B$ with $B = C_p^{p^{s-1}}$. A standard commutator calculation for semidirect products shows that $W_s' \cong W_{s-1}' \ltimes A$ where $A = (B \cap W_s')$ is generated by all $b(b^g)^{-1}$ with $b \in B$ and $g \in W_{s-1}$. Since W_{s-1} permutes the p^{s-1} copies of C_p in B transitively, A consists of those $(b_1,\ldots,b_{p^{s-1}}) \in B$ that satisfy $b_1 \cdots b_{p^{s-1}} = 1$. This implies that $B/A \cong C_p$ is cyclic of order p, and so W_s/W_s' is isomorphic to $W_{s-1}/W_{s-1}' \times C_p$. We deduce that $W_s/\Phi(W_s) \cong W_{s-1}/\Phi(W_{s-1}) \times C_p$ is elementary abelian of rank $(s-1)+1 = s$, so $\mathrm{rk}(W_s) = s$ by Corollary 1.8. □

Via σ_s, the group W_s acts uniserially on its module $T = (\mathbb{Z}_p^d, +)$ in dimension $d = d_s$. Let $T = T_0 > T_1 > \ldots$ be the associated uniserial series. For $j \in \mathbb{N}$, we denote the corresponding j-th two-step centraliser by

$$Z_j = C_{W_s}(T_j/T_{j+2}).$$

We now exhibit an alternative characterisation for these two-step centralisers. For this purpose we define the subgroups $M_1,\ldots,M_s$ of W_s using the generating set of

Proposition 4.22 via

$$M_i = \langle g_1, \ldots, g_{i-1}, g_{i+1}, \ldots, g_s\rangle \Phi(W_s).$$

By Propositions 1.6 and 4.22, each M_i is a maximal subgroup of W_s. Our proof of the following proposition provides the main connections between the two-step centralisers of W_s and the maximal subgroups $M_1, \ldots, M_s$. We cite Leedham-Green & McKay [57] for some of the required technical results.

Proposition 4.23 *The maximal subgroups $M_1, \ldots, M_s \leqslant W_s$ are exactly the distinct two-step centralisers of W_s in its action on $T = (\mathbb{Z}_p^{d_s}, +)$. If $s = 1$, then $M_1 = Z_i$ for all $i \in \mathbb{N}_0$. If $s \geqslant 2$, then the following hold.*

a) *$M_j = Z_i$ for $i \in \{kp - 1 \mid k \in p^{s-j-1}\mathbb{N}$ with $p^{s-j} \nmid k\}$ for $1 \leqslant j < s$.*

b) *$M_s = Z_i$ for all $i \in \mathbb{N}_0$ not covered by* a)*; that is, those with $p \nmid i + 1$.*

Proof Write $d = d_s$. Since $T_{i+d} = pT_i$ for all $i \in \mathbb{N}_0$, it follows that $Z_i = Z_{i+d}$ for each $i \in \mathbb{N}_0$; moreover, Proposition 4.18 asserts that $Z_i = Z_{i+p^{s-1}}$ for each $i \in \mathbb{N}_0$. Thus, the distinct two-step centralisers are contained in $\{Z_0, Z_1, \ldots, Z_{p^{s-1}-1}\}$.

Proposition 4.6 shows that each Z_i is a maximal subgroup of W_s, which implies that $\Phi(W_s) \leqslant Z_i$. Proposition 4.22 shows that $W_s/\Phi(W_s)$ is elementary abelian of order p^s, and $\{g_1\Phi(W_s), \ldots, g_s\Phi(W_s)\}$ is a minimal generating set. Thus the groups $M_1, \ldots, M_s$ are pairwise distinct. We now use induction on s to show that they are exactly the different two-step centralisers of W_s. For this purpose we extend our notation and write $M_{i,j} = \langle g_1, \ldots, g_{i-1}, g_{i+1}, \ldots, g_j\rangle\Phi(W_j)$ for $i \in \{1, \ldots, j\}$ and $j \in \mathbb{N}$. Hence $M_{i,s} = M_i$.

First, consider $s = 1$. In this case W_s has order p and $\{1\}$ is the only maximal subgroup. It follows that $M_{1,1} = Z_j$ for all $j \in \mathbb{N}_0$, as claimed.

Now let $s > 1$ and suppose the result holds for $s - 1$. We use the decompositions $W_s = C_p(g_s) \ltimes (B_1 \times \ldots \times B_p)$ and $T = D_1 \oplus \ldots \oplus D_p$ where each $B_i \cong W_{s-1}$ acts uniserially on $D_i \cong (\mathbb{Z}_p^{d/p}, +)$, and g_s permutes $B_1, \ldots, B_p$ and $D_1, \ldots, D_p$ cyclically. We assume that $B_1 = C_p(g_1) \wr \ldots \wr C_p(g_{s-1})$, so $W_s = B_1 \wr C_p(g_s)$. The induction hypothesis implies that $M_{1,s-1}, \ldots, M_{s-1,s-1}$ are distinct two-step centralisers in $B_1 = W_{s-1}$. Since $\Phi(B_1) = B_1'B_1^{[p]} \leqslant W_s'W_s^{[p]} = \Phi(W_s)$, it follows that

$$M_{i,s} = \langle M_{i,s-1}, g_s\rangle\Phi(W_s)$$

for each $i \in \{1, \ldots, s-1\}$. Since $M_{i,s-1}$ is a two-step centraliser for the action of B_1 on D_1, there exist $\mathbb{Z}_pB_1$-submodules $V \leqslant U \leqslant D_1$ with $[U : V] = p^2$ such that $M_{i,s-1}$ centralises U/V. Let $\varepsilon\colon D_1 \to D_1 \oplus \ldots \oplus D_p$, $u \mapsto (u, \ldots, u)$, be the diagonal embedding. Since $C_p(g_s)$ centralises $\varepsilon(D_1)$, it follows that $\varepsilon(V) < \varepsilon(U) \leqslant T$ are $\mathbb{Z}_pW_s$-submodules, and so $\varepsilon(U) = T_i$ and $\varepsilon(V) = T_{i+2}$ for some $i \in \mathbb{N}_0$, see Proposition 4.4. By construction, T_i/T_{i+1} is centralised by $M_{i,s-1}$ and by g_s. Moreover, $\Phi(W_s)$ lies in every 2-step centraliser, which allows us to deduce that T_i/T_{i+1} is centralised by $M_{i,s}$. In conclusion, $M_{1,s}, \ldots, M_{s-1,s}$ are two-step centralisers in W_s.

The group $M_{s,s}$ is the base group of $W_s = W_{s-1} \wr C_p$, hence $M_{s,s} = B_1 \times \ldots \times B_p$. For $k \in \mathbb{N}$ let $V \leqslant D_1$ be the unique $\mathbb{Z}_p B_1$-module of index p^k in D_1. Then $U = V \oplus \ldots \oplus V \leqslant D_1 \oplus \ldots \oplus D_p$ has index p^{kp}. Since U is $\mathbb{Z}_p W_s$-invariant, $U = T_{kp}$ by Proposition 4.4. Since $[B_i, V]$ has index p in V, it follows that $[M_{s,s}, U]$ has index p^p in U, hence $[M_{s,s}, U] = T_{(k+1)p}$. This implies that $M_{s,s} = Z_i$ for all i of the form $i = kp + j$ with $k \in \mathbb{N}_0$ and $j \in \{0, \ldots, p-2\}$.

Building on the first paragraph of the proof, it remains to show that the two-step centraliser Z_{kp-1} for $k = 1, \ldots, p^{s-2}$ are contained in $\{M_{1,s}, \ldots, M_{s,s}\}$. We refer to to [57, Lemma 4.2.3 and Proposition 10.3.2] for full details. The notation differs: for example, our Z_i is the $(i+1)$-th two step centraliser in [57, Proposition 10.3.2]. □

Corollary 4.24 *Let $A = \langle g_1 \cdots g_s \rangle \leqslant W_s$. Then A has order p^s and acts uniserially on $T = (\mathbb{Z}_p^{d_s}, +)$. Moreover, W_s has exponent p^s.*

Proof By Proposition 4.23, the group A is not contained in any two-step centraliser $M_1, \ldots, M_s$, hence it is uniserial by Proposition 4.6. Proposition 4.8 now shows that A is irreducible. Hence $|A| = p^s$ by Proposition 4.11. This implies that $p^s \mid \exp(W_s)$. We show equality using induction on s. Since $\exp(W_1) = p$, let $s > 1$ and assume that $\exp(W_{s-1}) = p^{s-1}$. If $h \in W_s = C_p \ltimes W_{s-1}^p$, then $h^p \in W_{s-1}^p$, and $|h^p|$ divides p^{s-1} by the induction hypothesis, that is, $\exp(W_s) \mid p^s$. □

Proposition 4.6 provides an abstract characterisation of the two-step centralisers of W_s in its action on T via σ_s. The following result shows that this characterisation also translates to the action of W_s on $(\mathbb{F}_p^{p^s}, +)$ via τ_s.

Lemma 4.25 *The action of W_s on $U = (\mathbb{F}_p^{p^s}, +)$ via τ_s is uniserial, and the subgroups $M_1, \ldots, M_s$ are exactly the distinct two-step centralisers of this action.*

Proof In the proof we identify W_s with $\tau_s(W_s)$. Corollary 4.24 shows that $g = g_1 \cdots g_s$ has order p^s, so it acts as the permutation matrix of a p^s-cycle. The minimal polynomial of g over $\mathbb{F}_p$ is $\mathbf{x}^{p^s} - 1 = (\mathbf{x} - 1)^{p^s}$ and the 1-eigenspace of g has dimension 1. By [45, Section 7.3], the matrix g is conjugate to its Jordan normal form

$$\begin{pmatrix} 1 & 1 & 0 & \ldots & 0 \\ 0 & 1 & 1 & & 0 \\ \vdots & & \ddots & \ddots & \vdots \\ 0 & 0 & \ldots & 1 & 1 \\ 0 & 0 & \ldots & 0 & 1 \end{pmatrix}.$$

Assuming that g has this form, a short calculation shows that the subgroup series $U = U_0 > U_1 > \ldots$ defined by $U_{i+1} = [\langle g \rangle, U_i]$ has length p^s, and so $\langle g \rangle$ acts uniserially on U by Proposition 4.2. We show that this series is W_s-invariant: since $W_s \ltimes U$ is a finite p-group, hence nilpotent. Thus $[W_s, W] < W$ for every submodule W of U. Clearly, U_0 is W_s-invariant. If U_i is W_s-invariant for some $i < p^s$, then $U_{i+1} = [\langle g \rangle, U_i] \leqslant [W_s, U_i] < U_i$, and $U_{i+1} = [W_s, U_i]$ follows from $[U_i : U_{i+1}] = p$. Now Proposition 4.2 implies that W_s acts uniserially on U, as claimed.

It remains to consider the two-step centralisers. For $s = 1$ the claim is trivially true. Let $s \geqslant 2$ and assume that the two-step centralisers of W_{s-1} via τ_{s-1} are as claimed. We first relate U to the W_s-module $T = (\mathbb{Z}_p^{d_s}, +)$ where W_s acts via σ_s. The constructions of σ_s and τ_s from ρ_s imply that T/pT is a quotient of U, say $T/pT = U/W$ for a submodule W of U of dimension $p^s - d_s = p^{s-1}$; thus, $W = U_{d_s}$. If we decompose $W_s = W_{s-1} \ltimes B$ with $B = C_p^{p^{s-1}}$, then the module W can also be described explicitly as in the construction of σ_s from ρ_s: specifically, W is spanned by p^{s-1} one-dimensional spaces, each corresponding to the 1-eigenspace $\langle(1, \ldots, 1)\rangle$ in the natural representation of one of the p^{s-1} subgroups C_p in B. In particular, W_s acts on this subspace W as W_{s-1} acts on $C_p^{p^{s-1}} \cong (\mathbb{F}_p^{p^{s-1}}, +)$ via τ_{s-1}.

Now consider the two-step centralisers. Since $pT = T_{d_s}$, for $i \in \mathbb{N}_0$ with $i+2 \leqslant d_s$, it follows that $\leqslant U_{i+2} < U_i$, and therefore

$$T_i/T_{i+2} \cong (T_i/pT)/(T_j/pT) \cong U_i/U_{i+2}$$

as W_s-modules. In this case, the two-step centraliser $C_{W_s}(U_i/U_{i+2}) = C_{W_s}(T_i/T_{i+2})$ is determined in Proposition 4.23. On the other hand, if $i \geqslant d_s$, then $U_{i+2} < U_i \leqslant W$, and the previous paragraph shows that $C_{W_s}(U_i/U_{i+2})$ can be determined by applying the induction hypothesis to W_{s-1} and τ_{s-1}.

It remains to determine $C = C_{W_s}(U_{d_s-1}/U_{d_s+1})$. Recall that $W_s = W_{s-1} \ltimes B$ where B is generated, as normal subgroup, by the generator g_1. By construction, W_{s-1} permutes the factors of B (and the underlying p-dimensional subspaces) via τ_{s-1}. Thus, as W_{s-1}-module, we can decompose U as a direct sum of p submodules $U^{(1)}, \ldots, U^{(p)}$ such that W_{s-1} acts via τ_{s-1} on each of them and $W = U_{d_s}$ is one of these direct summands, say $W = U^{(1)}$. Since $U_{d_s+1} < W < U_{d_s-1}$, there exists $w \in U^{(2)} \oplus \ldots \oplus U^{(p)}$ such that $U_{d_s-1} = W \oplus \langle w\rangle$. Since W and $\langle w\rangle$ are both W_{s-1}-invariant, the p-group W_{s-1} acts as a diagonal matrix on $U_{d_s-1}/U_{d_s+1} \cong C_p^2$, and therefore as the identity. This shows that $W_{s-1} \leqslant C$. Recall that W is the space of the fixed points of B in U, which is centralised by g_1. Thus, g_1 cannot centralise U_{d_s-1}/U_{d_s+1}, otherwise this quotient would be centralised by W_s, contradicting uniseriality. Thus, $g_1 \notin C$, and therefore C is a maximal subgroup of W_s that contains W_{s-1} and $\Phi(W_s)$, but not g_1. Therefore $C = M_1$ holds; see the discussion before Proposition 4.23. □

We now have all the information available to prove the following.

Proposition 4.26 *If $P \leqslant W_s$ acts uniserially on $T = (\mathbb{Z}_p^{d_s}, +)$, then* $\mathrm{rk}(P) \leqslant s$.

Proof Proposition 4.6 shows that $P \leqslant W_s$ is uniserial if and only if P is not contained in any of the subgroups $M_1, \ldots, M_s$ of W_s. We prove by induction on s that if the latter holds, then $\mathrm{rk}(P) \leqslant s$. As in the proof of Proposition 4.23, it is convenient to write $M_{1,s}, \ldots, M_{s,s}$.

If $s = 1$, then $W_1 \cong C_p$ and the claim is true. Now let $s > 1$ and assume that the claim is true for $s - 1$. As before, we consider $W_s = W_{s-1} \ltimes B$, where B is the base group of the wreath product and W_{s-1} acts uniserially on B via τ_{s-1}. Recall that B is generated as normal subgroup by g_1. In particular, $B \leqslant M_{2,s}, \ldots, M_{s,s}$

and $BM_{1,s} = W_s$. By construction, the natural homomorphism $W_s \to W_{s-1}$ with kernel B maps $M_{i,s}$ onto $M_{i-1,s-1}$ for $i = 2, \ldots, s$. If $P \leqslant W_s$ is not contained in any subgroup $M_{1,s}, \ldots, M_{s,s}$, then $PB/B \nleqslant M_{i-1,s-1}$ for $i = 2, \ldots, s$, and the induction hypothesis shows that $\mathrm{rk}(PB/B) \leqslant s - 1$.

Since $PB/B \cong P/P \cap B$, the claim follows if $P \cap B = \{1\}$. Now let $P \cap B \neq \{1\}$. Lemma 4.25 and our assumption that $P \nleqslant M_{1,s}, \ldots, M_{s,s}$ imply that $P \leqslant W_s$ acts uniserially via τ_s. In particular, P acts uniserially on B, and therefore on $P \cap B$. This implies that $P \cap B$ is generated, as a normal subgroup of P, by a single element, see Lemma 4.7. In conclusion, P is generated by at most $\mathrm{rk}(PB/B) + 1 \leqslant s$ elements, as claimed. □

4.3.2 The uniserial subgroups of V_s

The discussion of V_s is analogous to the previous case. In this section we consider $p = 2$; recall that $d_s = 2^{s-1}$ for $s \geqslant 3$. For $s = 3$, the group $V_3 = \langle a, b \rangle \leqslant \mathrm{GL}_4(\mathbb{Z}_2)$ is isomorphic to Q_{16}, see Proposition 4.12. If $s > 3$, then $V_s \leqslant \mathrm{GL}_{d_s}(\mathbb{Z}_2)$ is defined as $V_s = V_{s-1} \wr C_2 = C_2 \ltimes (V_{s-1} \times V_{s-1})$, where C_2 swaps the two copies of V_{s-1}. As before, we can also consider $V_s = Q_{16} \wr W_{s-3}$ where W_{s-3} is a Sylow 2-subgroup of $\mathrm{Sym}_{2^{s-3}}$ and permutes the 2^{s-3} direct factors Q_{16} accordingly. It follows that the order of V_s is

$$|V_s| = 16^{2^{s-3}} 2^{2^{s-4}+\ldots+2+1} = 2^{2^{s-1}+2^{s-3}-1}.$$

We denote $g_1 = ba$ and $g_2 = a$, and for $i \in \{3, \ldots, s-1\}$ we let g_i be a cyclic generator of the $(i-2)$-th factor C_2 in W_{s-3}; that is,

$$V_s = \langle g_1, g_2 \rangle \wr C_2(g_3) \wr \ldots \wr C_2(g_{s-1}),$$

which shows that V_s can be generated by $s - 1$ elements.

Proposition 4.27 *The group V_s has rank $s - 1$ for all $s \geqslant 3$.*

Proof Clearly, $\mathrm{rk}(V_3) = 2$, so let $s > 3$ and consider $V_s = W_{s-3} \ltimes B$ where B is a direct product of 2^{s-3} copies of Q_{16}. Analogous to the proof of Proposition 4.22, the group $A = B \cap V_s'$ satisfies $B/A \cong Q_{16}$, and so $V_s/V_s' \cong W_{s-3}/W_{s-3}' \times Q_{16}$. This implies $V_s/\Phi(V_s) \cong W_{s-3}/\Phi(W_{s-3}) \times C_2 \times C_2$, which has rank $s - 1$ by Proposition 4.22.□

The group V_s acts uniserially on $T = (\mathbb{Z}_2^d, +)$ in dimension $d = d_s = 2^{s-1}$, with series $T = T_0 > T_1 > \ldots$. The corresponding j-th two-step centraliser is

$$Z_j = C_{V_s}(T_j/T_{j+2})$$

for $j \in \mathbb{N}_0$. As in the case of odd primes, the distinct two-step centralisers are $M_1, \ldots, M_{s-1}$ where

$$M_i = \langle g_1, \ldots, g_{i-1}, g_{i+1}, \ldots, g_{s-1} \rangle \Phi(V_s).$$

The proof of the following result is similar to Proposition 4.23; we refer to [57, Proposition 10.3.2] for a detailed discussion.

Proposition 4.28 *The subgroups $M_1, \ldots, M_{s-1} \leqslant V_s$ are exactly the distinct two-step centralisers of V_s in its action on $T = (\mathbb{Z}_2^{d_s}, +)$. If $s = 3$, then $M_1 = Z_i$ for all odd i and $M_2 = Z_i$ for all even i. If $s > 3$, then the following hold.*

a) *$M_j = Z_i$ for $i \in \{2k-1 \mid k \in 2^{s-j-2}\mathbb{N}$ with $2^{s-j-1} \nmid k\}$ for $1 \leqslant j < s-1$, and*

b) *$M_{s-1} = Z_i$ for all $i \in \mathbb{N}_0$ not covered by* a)*; that is, even $i \in \mathbb{N}_0$.*

The proof of Corollary 4.24 now carries over to the analogous statement for $p = 2$.

Corollary 4.29 *Let $A = \langle g_1 \cdots g_{s-1} \rangle \leqslant V_s$. Then A has order 2^s and acts uniserially on $T = (\mathbb{Z}_2^{d_s}, +)$. Moreover, V_s has exponent 2^s.*

We conclude this section with a consideration of the rank of a uniserial subgroup of V_s, analogous to Proposition 4.26 for subgroups of W_s.

Proposition 4.30 *If $P \leqslant V_s$ acts uniserially on $T = (\mathbb{Z}_2^{d_s}, +)$, then* $\mathrm{rk}(P) \leqslant s + 1$.

Proof The claim is trivially true for $s = 3$ since every subgroup of $V_3 = Q_{16}$ has rank at most 2. Now consider $s > 3$ and decompose $V_s = W_{s-3} \ltimes B$ where $B = B_1 \times \ldots \times B_{2^{s-3}}$ with each $B_i \cong Q_{16}$. Each group B_i has a unique cyclic subgroup D_i of order 8, and we define $D = D_1 \times \ldots \times D_{2^{s-3}}$ as a subgroup of B. Now define $E_0 = B$, $E_1 = D$, $E_2 = D^{[2]}$, $E_3 = D^{[4]}$, and $E_4 = D^{[8]} = \{1\}$. Each of $E_0, \ldots, E_4$ is normal in V_s, and each quotient E_i/E_{i+1} with $i = 0, 1, 2, 3$ is elementary abelian of order 2^{s-3}. In particular, $B \leqslant C_{V_s}(E_i/E_{i+1})$ for all i. The group W_{s-3} acts uniserially on each E_i/E_{i+1} as described in Lemma 4.25. In particular, if $M \leqslant W_{s-3}$ is a two-step centraliser of the action on E_i/E_{i+1}, then the subgroup of V_s generated by M and B lies in some 2-step centraliser of V_s; this follows from Lemma 4.25 and Proposition 4.28.

Now let $P \leqslant V_s$ be a uniserial subgroup and consider PB/B as a subgroup of $W_{s-3} \cong V_s/B$. If PB/B lies in some two-step centraliser of W_{s-3} in its action on a quotient E_i/E_{i+1}, then the previous paragraph implies that P lies in some two-step centraliser of V_s, contradicting the uniseriality of P, see Proposition 4.6. Thus, PB/B is a uniserial subgroup of W_{s-3} (with respect to the action on a quotient E_i/E_{i+1}), and therefore $\mathrm{rk}(PB/B) \leqslant s - 3$ by Proposition 4.26. If $g_1, \ldots, g_{s-3} \in P$ project to a generating set of PB/B, then P is generated by these elements together with $P \cap B$. It remains to show that $P \cap B$ can be generated, as P-invariant subgroup, by at most four elements. Consider the series $B = E_0 > \ldots > E_4 = \{1\}$ of B. Since each $(P \cap E_i)E_{i+1}/E_{i+1}$ is a PB/B-invariant subspace of E_i/E_{i+1}, Lemma 4.7 implies that it is generated, as PB/B-module, by a single element. This holds for every $i = 0, 1, 2, 3$, so $\mathrm{rk}(P) \leqslant \mathrm{rk}(PB/B) + 4 \leqslant s + 1$, as claimed. □

Experimental evidence suggests that if $P \leqslant V_s$ is uniserial, then $\mathrm{rk}(P) \leqslant s - 1$.

4.4 Space groups with p-adic translation subgroup

Informally, a space group is the symmetry group of an object in a space. The algebraic characterisation of a space group says that a group G is a space group in dimension d if there exists a normal subgroup T of finite index in G such that T is self-centralising and a free module of rank d over the integers; that is, $T \cong (\mathbb{Z}^d, +)$. One of the surprising facts in coclass theory is that a p-adic variation of space groups plays a fundamental role here. We start with their formal definition.

Definition 4.31 A group G is a *p-adic space group* if there exists a normal subgroup T of p-power index in G such that T is self-centralising and a free module of finite rank over the p-adic integers, that is, $T = C_G(T)$ and $T \cong (\mathbb{Z}_p^d, +)$ for some $d \in \mathbb{N}$. The group T is a *translation subgroup* of G, and the finite p-group $P = G/T$ is the associated *point group*.

Proposition 4.32 *The translation subgroup of a p-adic space group is the unique maximal abelian normal subgroup.*

Proof Let T be a translation subgroup of G and let $S \trianglelefteq G$ be abelian. We show that $S \leqslant T$, which implies the claim. By Lemma 1.2, the subgroup ST is a nilpotent normal subgroup of G. Thus, $H_1 = T$ and $H_i = [ST, H_{i-1}]$ for $i \geqslant 2$ defines a series $T = H_1 > \ldots > H_{m+1} = \{1\}$ where each H_i/H_{i+1} is central in ST. On the other hand, T is a translation subgroup in G, and therefore $[ST : T]$ is finite and ST/T acts faithfully on T. Recall that $T \cong (\mathbb{Z}_p^d, +)$ for some $d \in \mathbb{N}$, and by choosing a $\mathbb{Z}_p$-basis of T associated to the central series constructed above, ST/T embeds faithfully as a finite subgroup into the group of upper unitriangular matrices in $\mathrm{GL}_d(\mathbb{Z}_p)$. Note that an upper triangular matrix $M \in \mathrm{GL}_d(\mathbb{Z}_p)$ of finite order k is the identity matrix: the minimal polynomial of such an M divides $(\mathbf{x} - 1)^d$ since M is unitriangular, and it also divides $\mathbf{x}^k - 1$ since M has order k. Thus, the minimal polynomial is $\gcd(\mathbf{x}^k - 1, (\mathbf{x} - 1)^d) = \mathbf{x} - 1$, and therefore M is the identity. This implies that ST/T is trivial, and so $S \leqslant T$ as claimed. □

Due to Proposition 4.32, the following is well-defined.

Definition 4.33 The *dimension* of a p-adic space group is the rank of its translation subgroup.

We now characterise those p-adic space groups that are infinite pro-p-groups of fixed coclass. For this purpose we investigate the action of a p-adic space group G on its translation subgroup T.

Definition 4.34 Let G be a p-adic space group with translation subgroup T and point group P. Then G is a *uniserial p-adic space group* if the action of P on T induced by conjugation is uniserial.

If G is a uniserial p-adic space group with translation subgroup T and point group P, then $T = C_G(T)$ is abelian, thus P acts uniserially and faithfully on T. If T has $\mathbb{Z}_p$-dimension d, then choosing a $\mathbb{Z}_p$-basis of T allows us to embed P as a subgroup of $\mathrm{GL}_d(\mathbb{Z}_p)$, and this is a finite p-group acting uniserially on $(\mathbb{Z}_p^d, +)$. Propositions 4.8 and 4.14 now show that the dimension d only takes special values.

Proposition 4.35 *The dimension of a p-adic uniserial space group is $d_s = (p-1)p^{s-1}$ for some $s \in \mathbb{N}$.*

Example 4.36 Consider the group $S = S(p^r)$ as constructed in Proposition 3.16: recall that $S = P \ltimes T$ where $P = \langle\theta\rangle$ is generated by a primitive p^r-th root of unity θ over $\mathbb{Q}_p$ that acts via multiplication on $T = (\mathbb{Z}_p[\theta], +)$. The group S is a uniserial p-adic space group of dimension $d = d_r$. Indeed, $T \cong (\mathbb{Z}_p^d, +)$ and $C_S(T) = T$, and the uniserial series $T = T_0 > T_1 > \ldots$ satisfies $T_i = (\mathfrak{p}^i, +)$, where $\mathfrak{p}$ is the ideal of $\mathbb{Z}_p[\theta]$ generated by $\theta - 1$. The properties mentioned in Proposition 4.4 already follow from the results in Section 3.4.

Next, we show that every uniserial p-adic space group G defines an infinite path in some coclass graph. Since G is finitely generated (as a topological group), we recall from Remark 1.21b) that the abstract subgroups $\gamma_i(G)$ are closed, so $\gamma_i(G) = \overline{\gamma_i(G)}$, and there is no ambiguity when we use $\gamma_i(G)$ below.

Proposition 4.37 *Let G be a uniserial p-adic space group with translation subgroup T, point group P of class k, and uniserial series $T = T_0 > T_1 > \ldots$.*

a) *There exists $\ell \in \{0, \ldots, k\}$ such that $\gamma_{k+1}(G) = T_\ell$.*

b) *If $\gamma_{k+1}(G) = T_\ell$ as in* a), *then* $\mathrm{cc}(G) = \mathrm{cc}(G/\gamma_{k+1}(G)) = \mathrm{cc}(P) + \ell$.

c) *The groups $G/\gamma_{k+1}(G), G/\gamma_{k+2}(G), \ldots$ form an infinite path in $\mathcal{G}(p, \mathrm{cc}(G))$.*

d) *If $G \cong P \ltimes T$, then $\ell = k$ and* $\mathrm{cc}(G) = \mathrm{cc}(P) + k = \log_p |P|$.

Proof a) Note that $T_0 = \gamma_1(G) \cap T$. Moreover, if $T_j = \gamma_i(G) \cap T$ for some $j < i$, then

$$T_{j+1} = [G, T_j] \leqslant \gamma_{i+1}(G) \cap T \leqslant \gamma_i(G) \cap T = T_j.$$

Since $[T_j : T_{j+1}] = p$, this shows $\gamma_{i+1}(G) \cap T = T_j$ or $\gamma_{i+1}(G) \cap T = T_{j+1}$. Thus, by induction, for every $i \in \mathbb{N}$ there exists some $\ell < i$ such that $\gamma_i(G) \cap T = T_\ell$. Since P has class k and $P \cong G/T$, it follows that $\gamma_{k+1}(G) \leqslant T$, so $\gamma_{k+1}(G) = \gamma_{k+1}(G) \cap T = T_\ell$ for some $\ell \in \{0, \ldots, k\}$, as claimed.

b) Since $\gamma_{k+1}(G) = T_\ell$ has index p^ℓ in T, uniseriality implies that

$$\mathrm{cc}(G) = \mathrm{cc}(G/\gamma_{k+1}(G)) = \log_p |G/\gamma_{k+1}(G)| - k = \log_p |P| + \ell - k,$$

and the latter equals $\mathrm{cc}(P) + \ell$.

c) If $\gamma_{k+1}(G) = T_\ell$, then, by uniseriality, $\gamma_{k+1+i}(G) = T_{\ell+i}$ and $\mathrm{cc}(G/\gamma_{k+1}(G)) = \mathrm{cc}(G/\gamma_{k+1+i}(G))$ for all $i \in \mathbb{N}_0$; the claim follows.

d) If G is such a split extension, then the inductive argument in a) shows that $\gamma_i(G) \cap T = T_{i-1}$ for $i \in \mathbb{N}$; the claim follows from b). □

4.5 Infinite pro-p-groups of finite coclass

Now we are in a position to present some structure results for the infinite pro-p-groups of finite coclass. The proof of the next theorem relies on Coclass Theorem C, which shows that an infinite pro-p-group of coclass r is solvable.

Theorem 4.38 *Let G be an infinite pro-p-group of coclass r.*

a) *There exists $i \in \mathbb{N}$ such that G acts uniserially on $\gamma_i(G)$.*

b) *The order of $Z(G)$ divides p^{r-1}, and $Z(G) \cap \gamma_i(G) = \{1\}$ for i as in* a).

c) *There exists $j \in \mathbb{N}$ such that $\gamma_j(G)$ is abelian.*

Proof a) Proposition 1.20 shows that there exists $i \in \mathbb{N}$ such that the index of $\gamma_{j+1}(G)$ in $\gamma_j(G)$ is p for all $j \geqslant i$. The proof of Coclass Theorem C in Section 1.4 shows that $\bigcap_{i\geqslant 2} \gamma_i(G) = \{1\}$. Thus, G acts uniserially on $\gamma_i(G)$, c.f. Definition 4.3. We note that if G is a uniserial p-adic space group, then the existence of i also follows from Proposition 4.37.

b) Let $U = Z(G) \cap \gamma_i(G)$ and suppose, for a contradiction, that $U \neq \{1\}$. Since $\bigcap_{\ell\geqslant 2} \gamma_\ell(G) = \{1\}$, there exists $k > i$ such that $U \leqslant \gamma_{k-1}(G)$ and $U \nleqslant \gamma_k(G)$, so $U\gamma_k(G)/\gamma_k(G)$ is a non-trivial central subgroup of $G/\gamma_k(G)$ contained in $\gamma_i(G)/\gamma_k(G)$. Since $G/\gamma_k(G)$ acts uniserially on $\gamma_i(G)/\gamma_k(G)$, Proposition 4.2 shows that $U\gamma_k(G)/\gamma_k(G) = \gamma_j(G/\gamma_k(G))$ for some $i \leqslant j < k$, so $U\gamma_k(G) = \gamma_j(G)$. Since U is central, $\gamma_{j+1}(G) = [G, U\gamma_k(G)] = [G, \gamma_k(G)] = \gamma_{k+1}(G)$, forcing $j = k$, a contradiction. This proves that $U = \{1\}$. Recall from Proposition 1.20 that G is the inverse limit of the groups $G/\gamma_j(G)$ for $j \geqslant i$. It follows that $G/Z(G)$ is the inverse limit of the groups $(G/Z(G))/\gamma_j(G/Z(G))$ for $j \geqslant i$. Since $Z(G) \cap \gamma_j(G) = \{1\}$ for every $j \geqslant i$, we observe that $\gamma_j(G/Z(G)) \cong \gamma_j(G)$, which has index $[G : \gamma_j(G)]p^{-z}$ in $G/Z(G)$. This proves that $G/Z(G)$ has coclass $r - z$, and so $z \leqslant r - 1$ by Corollary 1.11.

c) We follow the argument given in the proof of [19, Lemma 10.16]. Coclass Theorem C shows that G is solvable. Since G is infinite, there exists a unique $v \in \mathbb{N}$ such that the v-th term $V = G^{(v)}$ of the derived series of G has finite index in G and such that V/V' is infinite. In the following we employ some topological results listed in Remark 1.21: specifically, V is open in G and finitely generated (as a topological group), V' is closed in G, and so $V' = \bigcap_{k\geqslant 1} \gamma_k(G)V'$.

We show that $\gamma_j(G)V' > \gamma_{j+1}(G)V'$ for every $j \in \mathbb{N}$. If not, then there exists $j \in \mathbb{N}$ with $\gamma_j(G)V' = \gamma_{j+1}(G)V'$, and therefore

$$\gamma_j(G/V') = \gamma_j(G)V'/V' = \gamma_{j+1}(G)V'/V' = \gamma_{j+1}(G/V').$$

But then $\gamma_j(G/V') = \gamma_\ell(G/V')$ for every $\ell \geqslant j$, and so $V' = \bigcap_{k\geqslant 1} \gamma_k(G)V'$ equals $\gamma_j(G)V'$, that is, $\gamma_j(G) \leqslant V'$. Since $G/\gamma_j(G)$ is finite, G/V' is finite, which is impossible since V/V' is infinite. This contradiction shows that each quotient $Q_j = \gamma_j(G)V'/\gamma_{j+1}(G)V'$ is non-trivial.

The homomorphism $\gamma_\ell(G) \to Q_\ell$, $g \mapsto g\gamma_{\ell+1}(G)V'$, is surjective and has kernel $\gamma_\ell(G) \cap (\gamma_{\ell+1}(G)V')$. Dedekind's Modular Law [72, (1.3.14)] shows that this kernel equals $\gamma_{\ell+1}(G)(\gamma_\ell(G) \cap V')$. Thus $\gamma_\ell(G)/\gamma_{\ell+1}(G)(\gamma_\ell(G) \cap V') \cong Q_\ell$ is non-trivial. But $[\gamma_\ell(G) : \gamma_{\ell+1}(G)] = p$ for $\ell \geqslant i$, and therefore $\gamma_\ell(G) \cap V' \leqslant \gamma_{\ell+1}(G)$. This implies that $\gamma_\ell(G) \cap V' = \gamma_{\ell+1}(G) \cap V'$ for all $\ell \geqslant i$. Since $\bigcap_{\ell\geqslant 2} \gamma_\ell(G) = \{1\}$, we deduce that

$$\gamma_i(G) \cap V' = \bigcap_{\ell \geqslant i} (\gamma_\ell(G) \cap V') = \{1\}.$$

Since G/V is finite, there exists $j \geqslant i$ such that $\gamma_j(G) \leqslant V$. Now

$$\gamma_j(G)' \leqslant V' \cap \gamma_j(G) \leqslant V' \cap \gamma_i(G) = \{1\}$$

proves that $\gamma_j(G)$ is abelian, as claimed. □

We now present the main structure results.

Theorem 4.39 *An infinite pro-p-group G of coclass r either has trivial centre and is a uniserial p-adic space group, or has a central subgroup $N \leqslant Z(G)$ of order p such that G/N is an infinite pro-p-group of coclass $r - 1$.*

Proof We split the proof into various steps.

(1). We show that if G is a uniserial p-adic space group of coclass r, then $Z(G)$ is trivial. Let T be a translation subgroup of G, with point group P and uniserial series $T = T_0 > T_1 > \ldots$. Suppose, for a contradiction, that there exists a non-trivial $t \in Z(G)$. Propositions 4.4 and 4.32 show that $Z(G) \leqslant T$ and $\bigcap_{i \in \mathbb{N}} T_i = \{1\}$, so there exists $i \in \mathbb{N}_0$ such that $t \in T_i \setminus T_{i+1}$. In particular, the quotient T_i/T_{i+1} is generated by the image of t. Since $[G, T_{i+1}] = T_{i+2}$ and $[G, t] = 1$, this yields $[G, T_i] \leqslant T_{i+2}$, a contradiction. Thus, $Z(G) = \{1\}$ if G is a space group.

(2). We show that if G is an infinite pro-p-group of coclass r with non-trivial centre, then there exists a central subgroup N of order p such that G/N is an infinite pro-p-group of coclass $r - 1$. If $Z(G)$ is non-trivial, then it is a non-trivial finite p-group by Theorem 4.38, and so G has a normal subgroup N of order p by Proposition 1.4. Now Theorem 4.38b) and its proof imply that there exists $i \in \mathbb{N}_0$ such that for every $j \geqslant i$ it follows that $N \cap \gamma_j(G) = \{1\}$ and $[G/N : \gamma_j(G/N)] = [G : \gamma_j(G)]p^{-1}$; in particular, G/N is an infinite pro-p-group of coclass $r - 1$, as claimed.

(3). We show that an infinite pro-p-group G of coclass r with trivial centre is a uniserial space group. Since G has coclass r, there exists $i \in \mathbb{N}$ such that G acts uniserially on $\gamma_i(G)$; by Theorem 4.38, we can assume that $\gamma_i(G)$ is abelian. Since $\gamma_i(G)$ has finite index in G, this also shows that we can choose a non-trivial normal subgroup T of G that is maximal with respect to being abelian and of finite index in G. It follows from Remark 1.21 that T is finitely generated as a topological group. The following steps show that G is a uniserial space group.

(3a) Free $\mathbb{Z}_p$-module. We show that T is a free $\mathbb{Z}_p$-module of finite rank. Let U be the torsion subgroup of T, that is, the subgroup of T generated by all elements of finite order. Since U is characteristic in T, it is normal in G. Since T is abelian and finitely generated as a topological group, U is a finite pro-p-group, hence a finite p-group, on which G/T acts by conjugation. If $U \neq \{1\}$, then Proposition 1.4 shows that there is a non-trivial $u \in U$ that is fixed by G/T; but then $u \in Z(G)$, which is not possible since $Z(G) = \{1\}$ by assumption. This contradiction shows that $U = \{1\}$. It follows that T is a free $\mathbb{Z}_p$-module of finite rank.

(3b) Self-centralising. Let $C = C_G(T)$; we show that $T = C$. Since T is normal in G, it follows that $C \trianglelefteq G$. If $C > T$, then C/T is a non-trivial normal subgroup of the finite p-group $P = G/T$, and Proposition 1.4 shows that there exists $c \in C \setminus T$ such that $cT \in Z(G/T)$; now the topological group generated by T and c is abelian, normal, and of finite index (hence open) in G (see Remark 1.21), which contradicts the maximality of T. This proves that $C = T$, and therefore P acts faithfully on T.

(3c) Uniserial. It remains to show that P acts uniserially on T; recall that P acts uniserially on $\gamma_i(G)$. Since G/T is finite, there exists $j \geqslant i$ such that $\gamma_j(G) \leqslant T$ and P acts uniserially on $\gamma_j(G)$. Since $T/\gamma_j(G)$ is a finite p-group, there exists some p-power q such that $T^{[q]} \leqslant \gamma_j(G)$. But $T/T^{[q]}$ is also finite, so we can choose $k \geqslant j$ with $\gamma_k(G) \leqslant T^{[q]}$. Now $G/\gamma_k(G)$ acts uniserially on $\gamma_j(G)/\gamma_k(G)$, and $T^{[q]}/\gamma_k(G)$ is a G-invariant subgroup, so $T^{[q]}/\gamma_k(G) = \gamma_\ell(G)/\gamma_k(G)$ for some $k \leqslant \ell \leqslant j$ by Proposition 4.4. Thus, $T^{[q]} = \gamma_\ell(G)$, and G acts uniserially on $T^{[q]}$. Since T is abelian, basic commutator identities show that if $t \in T$ and $g \in P$, then $[g, t^q] = [g, t]^q$. Thus exponentiation by q defines an isomorphism $T \longrightarrow T^{[q]}$ that is compatible with the action of G. Since G acts uniserially on $T^{[q]}$, it also acts uniserially on T, as claimed.

Concluding Case (3), it follows that G is a uniserial p-adic space group with translation subgroup T and point group P. □

The *hypercentre* of G is the union of the terms of the upper central series of G.

Corollary 4.40 *Let G be an infinite pro-p-group of coclass r with hypercentre H.*

a) *The group H has order p^k for some $k \leqslant r-1$, and G/H is a uniserial p-adic space group of coclass $r-k$.*

b) *There exists $\ell \geqslant 0$ such that $\gamma_\ell(G)$ is a free $\mathbb{Z}_p$-module with $\gamma_\ell(G) \cap H = \{1\}$, and $G/\gamma_\ell(G)$ is a finite p-group of coclass r that acts uniserially on $\gamma_\ell(G)$.*

c) *Part* b) *holds for $\ell > p^{r-1}$.*

Proof The first claim is immediate from Theorem 4.39, which also shows that G/H is a uniserial p-adic space group (with trivial centre), say of dimension d. Part (3c) of the proof of Theorem 4.39 shows that there exists $\ell \geqslant 0$ such that G acts uniserially on the abelian group $\gamma_\ell(G)$, which is a free $\mathbb{Z}_p$-module of dimension d, and $G/\gamma_\ell(G)$ is a finite p-group. By Theorem 4.38, we can assume that ℓ is large enough such that H intersects $\gamma_\ell(G)$ trivially. The last claim follows from [57, Theorem 11.2.4], which shows that the results hold for all $\ell > p^{r-1}$. □

An infinite pro-p-group G of coclass r with non-trivial centre is also called a uniserial *pre-space group* with translation subgroup $T = \gamma_\ell(G)$ and point group $P = G/T$. Note that if G is not a space group, then the translation subgroup is not necessarily unique and P does not necessarily act faithfully on T.

4.6 Application: The class-breadth conjecture

Let G be a finite p-group of class $\mathrm{cl}(G)$. By the Orbit-Stabiliser Theorem, the size of every conjugacy class of elements of G is a p-power. The *breadth* of G is the largest integer $\mathrm{b}(G)$ such that $p^{\mathrm{b}(G)}$ is the size of a class. Leedham-Green, Neumann

& Wiegold [59] proved that $\mathrm{cl}(G) \leqslant \mathrm{b}(G)p/(p-1)+1$, and they formulated the *Class-Breadth Conjecture* claiming that

$$\mathrm{cl}(G) \leqslant \mathrm{b}(G) + 1.$$

While this conjecture was proved for many families of groups, it is not true in general: Felsch, Neubüser & Plesken [37] used the theory of uniserial 2-adic space groups to construct finite 2-groups that violated the conjecture. More counterexamples are described by Eick, Newman & O'Brien [35] who also investigated the breadths of quotients of p-adic space groups. We briefly recall this application of the theory of infinite pro-p-groups of finite coclass.

For a finite p-group G of order p^n and breadth $\mathrm{b}(G)$, the *cobreadth* is defined as $\mathrm{cb}(G) = n - \mathrm{b}(G)$. The class-breadth conjecture can now be formulated as

$$\mathrm{cb}(G) \leqslant \mathrm{cc}(G) + 1.$$

Let S be an infinite pro-p-group of finite coclass r. Recall that there exists $j \in \mathbb{N}$ such that $\mathrm{cc}(S/\gamma_i(S)) = r$ for all $i \geqslant j$. The cobreadths $\mathrm{cb}(S/\gamma_i(S))$ for $i \geqslant j$ form a non-decreasing sequence of integers, see [37, p. 115]. The following result is obtained in [35, Corollary 2.4] and allows us to define $\mathrm{cb}(S) = \lim_{i\to\infty} \mathrm{cb}(S/\gamma_i(S))$.

Lemma 4.41 *Let S be a uniserial p-adic space group. Then there exists $b \in \mathbb{N}$ with $\mathrm{cb}(S/\gamma_i(S)) \leqslant b$ for all $i \in \mathbb{N}$.*

Let T be a translation subgroup of S with point group P and uniserial series $T = T_0 > T_1 > \ldots$. Following the notation in [35], the group S is *covered* if P is the set-wise union of the two-step centralisers of S; that is, if

$$P = \bigcup_{i \in \mathbb{N}_0} C_P(T_i/T_{i+2}).$$

By Corollary 4.15, the group P is conjugate to a subgroup of W_s or of V_s, and the two-step centralisers of the latter groups were determined in Propositions 4.23 and 4.28 as the maximal subgroups $M_1, \ldots, M_s \leqslant W_s$ and $M_1, \ldots, M_{s-1} \leqslant V_s$, respectively. If $P \leqslant W_s$ is uniserial, then P and W_s induce the same uniserial series, and therefore the two-step centralisers of P are $P \cap M_1, \ldots, P \cap M_s$; analogously for $P \leqslant V_s$. Thus, the property of being covered can be read off from the intersections of P with the groups M_i. The following result is [35, Lemma 2.5].

Theorem 4.42 *If S is an uncovered uniserial p-adic space group, then S satisfies the class-breadth conjecture, that is,* $\mathrm{cb}(S) \leqslant \mathrm{cc}(S) + 1$. *In particular, there exists $j \in \mathbb{N}$ such that $S/\gamma_i(S)$ satisfies the class-breadth conjecture for all $i \geqslant j$.*

An immediate consequence is the following.

Corollary 4.43 *Let S be a uniserial p-adic space group in dimension d_s. If the point group of S has exponent p^s, then S is uncovered and therefore satisfies the class-breadth conjecture.*

Proof First, let p be odd. Up to conjugacy, we can assume that the point group of S is a subgroup of W_s. The proof of Corollary 4.24 implies that an element of order p^s has the form $g_1^{e_1} \dots g_s^{e_s}$ with each $e_i \in \{1, \dots, p-1\}$. In particular, this element is not contained in any two-step centraliser, so S is uncovered. Now Theorem 4.42 proves the claim. The proof for $p = 2$ is similar and uses Corollary 4.29. □

Counterexamples to the class-breadth conjecture were obtained in [35, 37] by constructing covered uniserial p-adic space groups and checking their cobreadth. For $p = 3$, covered space groups occur at dimension 54, and it was proved in [35, Theorem 5.1] that these groups satisfy the class-breadth conjecture. For $p = 2$, covered space groups arise in dimension 4. There are eight such groups and exactly one gives rise to counterexamples: the quotient $G/\gamma_{16}(G)$ of the group

$$G = \langle a, b \mid a^4, b^4, [[b, a], a], [b^2, a]^2 \rangle$$

has order 2^{19}, coclass 4, and breadth 6. We refer to [35, Section 4.1] for details and references. The class-breadth conjecture remains open for odd primes.

Chapter 5
Cohomology theory and extensions

The cohomology of finite p-groups plays a crucial role in coclass theory. For instance, the proof of Periodicity I presented in Chapter 7 relies fundamentally on cohomological methods. In this chapter we develop the basic theory of group cohomology from first principles, and build up to the central results required for the applications that follow. Although the material presented here is mostly classical, we include it to provide a self-contained and coherent account.

5.1 Basic definitions

This section recalls the basic concepts of group cohomology. While these may be familiar to many readers, we provide a concise overview to maintain coherence in the presentation. For a more comprehensive introduction to cohomology, we refer the reader to [72], as well as [4, 44, 79].

5.1.1 Cohomology groups

Throughout, let G be a group and let M be a G-module, that is, an abelian group with a G-action. We usually write M additively and denote the action of $g \in G$ on $m \in M$ by m^g. As before, for $n \in \mathbb{N}$ we denote by G^n the direct product of n copies of G.

The group of n-th *cochain maps* $C^n(G, M)$ is defined via $C^0(G, M) = M$ and

$$C^n(G, M) = \{\varphi\colon G^n \to M \mid \varphi(g_1, \ldots, g_n) = 0 \text{ if some } g_i = 1\}$$

for $n \in \mathbb{N}$. If $\varphi, \psi \in C^n(G, M)$, then setting $(\varphi + \psi)(g) = \varphi(g) + \psi(g)$ for $g \in G^n$ furnishes $C^n(G, M)$ with the structure of an abelian group.

H. Dietrich and B. Eick, *Coclass graphs of p-groups*, Springer Monographs in Mathematics, https://doi.org/10.1007/978-3-032-17506-9_5

If $n \in \mathbb{N}$, then the n-th *cohomology map* is the function

$$\delta_n\colon C^{n-1}(G, M) \to C^n(G, M)$$

that takes $\varphi \in C^{n-1}(G, M)$ to the map $\delta_n(\varphi) \in C^n(G, M)$ defined by

$$\begin{aligned}\delta_n(\varphi)(g_1, \ldots, g_n) = {} & \varphi(g_2, \ldots, g_n) \\ & + \sum_{i=1}^{n-1} (-1)^i \varphi(g_1, \ldots, g_{i-1}, g_i g_{i+1}, \ldots, g_n) \\ & + (-1)^n (\varphi(g_1, \ldots, g_{n-1}))^{g_n}.\end{aligned}$$

The map δ_n is a group homomorphism. Note that if $n = 1$ and $m \in C^0(G, M)$, then $\delta_1(m)$ is the map that takes $g \in G$ to $\delta_1(m)(g) = m - m^g$. If $n = 2$ and $\varphi \in C^1(G, M)$, then $\delta_2(\varphi)(g, h) = \varphi(h) - \varphi(gh) + \varphi(g)^h$. In particular, $\delta_2(\delta_1(m))$ is the 0-map $G^2 \to M$. This is true in general: a calculation shows that the composition $\delta_{n+1} \circ \delta_n$ is the 0-map for every $n \in \mathbb{N}$, that is, $\operatorname{im} \delta_n \leqslant \ker \delta_{n+1}$. This property is used to define cohomology groups.

Definition 5.1 Let G be a group and M a G-module. Define $B^0(G, M) = \{0\}$ and for $n \in \mathbb{N}$ let

$$Z^{n-1}(G, M) = \ker \delta_n \quad \text{and} \quad B^n(G, M) = \operatorname{im} \delta_n.$$

Then $Z^n(G, M)$ is the group of n-th *cocycles* and $B^n(G, M)$ are the n-th *coboundaries*. The n-th *cohomology group* is the quotient group

$$H^n(G, M) = Z^n(G, M)/B^n(G, M).$$

For example, $Z^2(G, M)$ consists of all $\gamma \in C^2(G, M)$ that satisfy

$$\gamma(gh, k) + \gamma(g, h)^k = \gamma(g, hk) + \gamma(h, k) \quad \text{(for all } g, h, k \in G), \tag{5.1}$$

and the subgroup $B^2(G, M)$ consists of all $\gamma \in Z^2(G, M)$ for which there exists a map $\varphi\colon G \to M$ such that $\gamma(g, h) = -\varphi(gh) + \varphi(g)^h + \varphi(h)$ for all $g, h \in G$.

We collect some results for later applications. Recall that the *exponent* of a multiplicative group H, denoted $\exp(H)$, is the smallest integer $n > 0$ such that $t^n = 1$ for every $t \in H$, and $\exp(H) = \infty$ if such an n does not exist. Recall that an additive abelian group M is *uniquely divisible* by $m \in \mathbb{N}$ if for every $a \in M$ there exists a unique $b \in M$ with $a = mb$.

Proposition 5.2 *Let G be a finite group, M a G-module, and $n \in \mathbb{N}$.*

a) *The exponent* $\exp(H^n(G, M))$ *divides* $|G|$.

b) *If M is uniquely divisible by $|G|$, then $H^n(G, M) = \{0\}$.*

We note that for both finite and infinite groups G, the values $\exp(Z^n(G, M))$ and $\exp(H^n(G, M))$ divide $|M|$.

The cohomology of cyclic groups can be determined explicitly.

Proposition 5.3 *Let $G = \langle g \rangle$ be finite, let M be a G-module, and let $n \in \mathbb{N}$. Define $\alpha\colon M \to M$, $a \mapsto a^g - a$, and $\beta\colon M \to M$, $a \mapsto a + a^g + \ldots + a^{g^{|G|-1}}$. Then*

$$H^{2n-1}(G, M) \cong \ker\beta/\operatorname{im}\alpha \quad \textit{and} \quad H^{2n}(G, M) \cong \ker\alpha/\operatorname{im}\beta.$$

5.1.2 Group extensions and equivalence

Throughout this subsection, let G be a group and let M be an abelian group, written additively. An extension of G by M is a group E that has a normal subgroup $N \cong M$ such that $E/N \cong G$. Equivalently, E is an extension of G by M if there exists a short exact sequence of group homomorphisms

$$\{0\} \longrightarrow M \stackrel{\iota}{\longrightarrow} E \stackrel{\pi}{\longrightarrow} G \longrightarrow \{1\},$$

that is, ι is injective, π is surjective, and $\ker\pi = \operatorname{im}\iota$.

Remark 5.4 One usually identifies M with $\iota(M)$. However, we introduced M as an additive group (to facilitate working with modules later), while the group operations of the possibly non-abelian group E and its normal subgroup $\iota(M)$ are typically written multiplicatively. In this section only, we distinguish between M and $\iota(M)$, and use the isomorphism $M \to \iota(M)$ induced by ι to switch between the additive and multiplicative group.

An extension E of G by M defines a G-module structure on M and a 2-cocycle in $Z^2(G, M)$ as follows. For each $g \in G$ choose $\tau(g) \in E$ in the full preimage of g under π. We assume that $\tau(1) = 1$. Since $\iota(M)$ is the kernel of π, the set $\{\tau(g) \mid g \in G\}$ is a transversal for $\iota(M)$ in E. Hence every $h \in E$ has a unique factorisation $h = \tau(g)\iota(m)$ for some $g \in G$ and $m \in M$. By abuse of notation, we also call $\tau\colon G \to E$ a transversal. It follows that $\pi(\tau(g))\pi(\tau(h)) = gh = \pi(\tau(gh))$ for all $g, h \in G$. Thus there exists a unique $\gamma(g, h) \in M$ such that

$$\tau(g)\tau(h) = \tau(gh)\iota(\gamma(g, h)).$$

This defines a map $\gamma\colon G \times G \to M$; as $\tau(1) = 1$, it satisfies $\gamma(1, g) = \gamma(g, 1) = 0$ for all $g \in G$. In fact, γ is a 2-cocycle as shown in the next result.

Lemma 5.5 *With the previous notation we obtain the following.*

a) *The group G acts on $\iota(M)$ via $\iota(m)^g = \iota(m)^{\tau(g)}$ for $g \in G$ and $m \in M$, where the latter is conjugation in E; this induces a G-module action on M via $m^g = \iota^{-1}(\iota(m)^g)$.*

b) *With the G-module structure on M as in* a), *the map $\gamma\colon G \times G \to M$ is a 2-cocycle in $Z^2(G, M)$.*

c) *The underlying set of E is $\{\tau(g)\iota(m) \mid g \in G, m \in M\}$ with multiplication*

$$\begin{aligned} \tau(g)\iota(m)\tau(h)\iota(n) &= \tau(gh)\iota(m)^h\iota(n)\iota(\gamma(g, h)) \\ &= \tau(gh)\iota(m^h + n + \gamma(g, h)). \end{aligned} \tag{5.2}$$

Proof a) Observe $\tau(1) = 1$, so $\iota(m)^1 = \iota(m)^{\tau(1)} = \iota(m)$. As $\iota(M)$ is abelian, if $m \in M$ and $g, h \in G$, then

$$\begin{aligned}\iota(m)^{gh} &= \iota(m)^{\tau(gh)} \\ &= \iota(m)^{\tau(g)\tau(h)\iota(\gamma(g,h))} \\ &= \iota(m)^{\tau(g)\tau(h)} = (\iota(m)^g)^h.\end{aligned}$$

Clearly, $\iota(m) \mapsto \iota(m)^g$ is an automorphism of $\iota(M)$; the claim follows.

b) Explicitly computing $(\tau(g)\tau(h))\tau(k)$ and $\tau(g)(\tau(h)\tau(k))$ for $g, h, k \in G$, the associativity of E implies that (5.1) holds, so $\gamma \in Z^2(G, M)$.

c) This follows directly. □

The construction of γ depends on the chosen transversal $\tau: G \to E$. What happens if we choose a different transversal, say $\sigma: G \to E$ with $\sigma(1) = 1$? Since $\pi(\sigma(g)) = \pi(\tau(g)) = g$ for all $g \in G$, there is a function $\kappa: G \to M$ such that $\sigma(g) = \tau(g)\iota(\kappa(g))$ for all $g \in G$; note that $\kappa(1) = 0$, so $\kappa \in C^1(G, M)$. This implies the following. First, the G-module structure defined by σ is that defined by τ. Second, the 2-cocycle δ defined by σ satisfies the following for all $g, h \in G$:

$$\begin{aligned}\iota(\delta(g,h)) &= \sigma(gh)^{-1}\sigma(g)\sigma(h) \\ &= (\tau(gh)\iota(\kappa(gh)))^{-1}\tau(g)\iota(\kappa(g))\tau(h)\iota(\kappa(h)) \\ &= \iota(\kappa(gh))^{-1}\tau(gh)^{-1}\tau(g)\tau(h)\iota(\kappa(g))^h\iota(\kappa(h)) \\ &= \iota(\gamma(g,h))\iota(\kappa(gh))^{-1}\iota(\kappa(g))^h\iota(\kappa(h)),\end{aligned}$$

and therefore

$$\delta(g,h) = \gamma(g,h) - \kappa(gh) + \kappa(g)^h + \kappa(h).$$

Note that $-\kappa(gh) + \kappa(g)^h + \kappa(h)$ is the image of $(g, h) \in G \times G$ under the 2-coboundary $\delta_2(\kappa) \in B^2(G, M)$. It follows that δ and γ differ by the 2-coboundary $\delta_2(\kappa) \in B^2(G, M)$, which proves the following.

Lemma 5.6 *Two 2-cocycles arising from two different transversals to M in an extension E are in the same cohomology class in $H^2(G, M)$.*

We have seen how every extension of G by an abelian group M defines a G-module structure on M and a 2-cocycle $\gamma \in Z^2(G, M)$, with respect to a chosen transversal. We now reverse this process and describe how to construct an extension from a group, module, and 2-cocycle.

Definition 5.7 Let G be a group and let M be a G-module. The extension of G by M defined by a 2-cocycle $\gamma \in Z^2(G, M)$ is the set $E_\gamma = G \times M$ equipped with the multiplication $(g, m)(h, n) = (gh, m^h + n + \gamma(g, h))$.

Note that E_γ is a group: associativity follows from the 2-cocycle condition (5.1), the identity is $(1, 0)$, and the inverse of (g, n) is $(g, n)^{-1} = (g^{-1}, -n^{g^{-1}} - \gamma(g, g^{-1}))$. Further, E_γ is an extension of G by M: the subgroup $\{1\} \times M$ is normal in E_γ, and its quotient is isomorphic to G. The map $\iota\colon M \to E_\gamma$, $m \mapsto (1, m)$, is an embedding, and $\tau\colon G \to E_\gamma$, $g \mapsto (g, 0)$, is a transversal to $\iota(M)$ in E_γ.

Since $\tau(gh)^{-1}\tau(g)\tau(h) = (1, \gamma(g, h))$ for all $g, h \in G$, the 2-cocycle defined by τ is γ; for this calculation the identity $\gamma(k^{-1}, k) = \gamma(k, k^{-1})^k$ is useful (it holds by (5.1)). We also note that the conjugation action of E_γ on M is defined by the module action: if $g \in G$ and $m, n \in M$, then $(1, m)^{(g,n)} = (1, m^g)$. The following result is now a direct consequence of (5.2).

Proposition 5.8 *Let $\{0\} \to M \xrightarrow{\iota} E \xrightarrow{\pi} G \to \{1\}$ be an extension, and let γ in $Z^2(G, M)$ be the 2-cocycle defined by a chosen transversal $\tau\colon G \to E$ and the induced G-module structure on M. Then E and E_γ are isomorphic, and an isomorphism is given by $E \to E_\gamma$, $\tau(g)\iota(m) \mapsto (g, m)$.*

Two extensions E and $\tilde{E}$ of G by M, given by short exact sequences, are *equivalent extensions* if there is an isomorphism $\xi\colon E \to \tilde{E}$ such that the following diagram is commutative.

$$\begin{array}{ccccccccc}
\{0\} & \longrightarrow & M & \xrightarrow{\iota} & E & \xrightarrow{\pi} & G & \longrightarrow & \{1\} \\
 & & \| & & \downarrow{\scriptstyle \xi} & & \| & & \\
\{0\} & \longrightarrow & M & \xrightarrow{\tilde{\iota}} & \tilde{E} & \xrightarrow{\tilde{\pi}} & G & \longrightarrow & \{1\}
\end{array}$$

The isomorphism in Proposition 5.8 shows that every extension E of G by M is equivalent to an extension of the form E_γ for some 2-cocycle $\gamma \in Z^2(G, M)$, with respect to the induced G-module structure of M. Moreover, the following holds.

Proposition 5.9 *Let M be a G-module and $\gamma, \delta \in Z^2(G, M)$. The extensions E_γ and E_δ are equivalent if and only if γ and δ are in the same cohomology class in $H^2(G, M)$.*

Proof Let $\xi\colon E_\gamma \to E_\delta$ be an isomorphism that proves equivalence. This implies that there is a map $\kappa \in C^1(G, M)$ such that the image of each $(g, m) \in E_\gamma$ under ξ has the form $(g, m + \kappa(g))$. Now the group homomorphism property implies that $\gamma = \delta + \delta_2(\kappa)$ with $\delta_2(\kappa) \in B^2(G, M)$, as claimed. Conversely, a similar calculation shows that if $\gamma = \delta + \delta_2(\kappa)$ for some $\kappa \in C^1(G, M)$, then the required isomorphism $E_\gamma \to E_\delta$ can be defined as $(g, m) \mapsto (g, m + \kappa(g))$. □

Remark 5.10 Let G be a group and let M be a G-module. If $\gamma \in H^2(G, M)$ is a cohomology class, then (by abuse of notation) we write E_γ for the extension of G by M defined by a 2-cocycle in the class γ, as explained in Definition 5.7. By Proposition 5.9, this definition of E_γ is unique up to equivalence of extensions.

5.1.3 Compatible pairs and strong isomorphism

Associated with the G-module structure in M is the group of compatible pairs. Their impact on the isomorphism problem for group extensions is well known, see for example [46, Section 8.9].

Definition 5.11 Let G be a group and M a G-module. For $g \in G$ denote by $\overline{g} \in \mathrm{Aut}(M)$ the automorphism $M \to M$, $m \mapsto m^g$. The group $\mathrm{Comp}(G, M)$ of *compatible pairs* is

$$\{(\alpha, \beta) \in \mathrm{Aut}(G) \times \mathrm{Aut}(M) \mid \overline{\alpha(g)} = \beta \circ \overline{g} \circ \beta^{-1} \text{ for all } g \in G\}.$$

Note that $\alpha \in \mathrm{Aut}(G)$ and $\beta \in \mathrm{Aut}(M)$ satisfy $(\alpha, \beta) \in \mathrm{Comp}(G, M)$ if and only if $\beta(m^g) = \beta(m)^{\alpha(g)}$ for all $g \in G$ and $m \in M$. The set of compatible pairs is a subgroup of the direct product $\mathrm{Aut}(G) \times \mathrm{Aut}(M)$.

The group $\mathrm{Comp}(G, M)$ acts on $C^n(G, M)$ from the left for every $n \in \mathbb{N}$; if $(\alpha, \beta) \in \mathrm{Comp}(G, M)$ and $\varphi \in C^n(G, M)$, then $(\alpha, \beta)(\varphi) \in C^n(G, M)$ is the map defined by

$$(\alpha, \beta)(\varphi)(g_1, \ldots, g_n) = \beta(\varphi(\alpha^{-1}(g_1), \ldots, \alpha^{-1}(g_n))). \tag{5.3}$$

This action leaves $Z^n(G, M)$ and $B^n(G, M)$ setwise invariant and thus allows us to define an induced action on $H^n(G, M)$, where $(\alpha, \beta) \in \mathrm{Comp}(G, M)$ maps the cohomology class $\varphi + B^n(G, M)$ to $(\alpha, \beta)(\varphi) + B^n(G, M)$.

Definition 5.12 Two extensions E and $\tilde{E}$ of G by M, given by short exact sequences, are *strongly isomorphic* if there are automorphisms $\beta \in \mathrm{Aut}(M)$, $\alpha \in \mathrm{Aut}(G)$, and an isomorphism $\xi \colon E \to \tilde{E}$ such that the following diagram is commutative.

$$\begin{array}{ccccccccc}
\{0\} & \longrightarrow & M & \xrightarrow{\iota} & E & \xrightarrow{\pi} & G & \longrightarrow & \{1\} \\
 & & \downarrow{\scriptstyle \beta} & & \downarrow{\scriptstyle \xi} & & \downarrow{\scriptstyle \alpha} & & \\
\{0\} & \longrightarrow & M & \xrightarrow{\tilde{\iota}} & \tilde{E} & \xrightarrow{\tilde{\pi}} & G & \longrightarrow & \{1\}
\end{array}$$

Equivalent extensions are strongly isomorphic, but the converse does not hold. We now discuss how compatible pairs can be used to investigate strong isomorphisms and automorphisms.

Proposition 5.13 *Let M be a G-module and $\gamma, \delta \in Z^2(G, M)$. The extensions E_γ and E_δ are strongly isomorphic if and only if the cohomology classes defined by γ and δ are in the same orbit under the action of* $\mathrm{Comp}(G, M)$.

Proof If $E = E_\gamma$ and $\tilde{E} = E_\delta$ are strongly isomorphic with the maps defined in the diagram of Definition 5.12, then there exists a map $\kappa \in C^1(G, M)$ such that the image of each $(g, m) \in E_\gamma$ under ξ has the form $(\alpha(g), \beta(m) + \kappa(\alpha(g)))$. The group homomorphism property $\xi((g, m)(h, n)) = \xi((g, m))\xi((h, n))$ for all $g, h \in G$ and $n, m \in M$ implies that

$$(*)\qquad \beta(\gamma(g,h)) + \beta(m^h) + \kappa(\alpha(gh)) + \beta(n) = \delta(\alpha(g),\alpha(h)) + \beta(m)^{\alpha(h)} + \kappa(\alpha(g))^{\alpha(h)} + \kappa(\alpha(h)) + \beta(n).$$

Setting $g = 1$ and $n = 0$ shows that $\beta(m^h) = \beta(m)^{\alpha(h)}$ for all $h \in G$ and $m \in M$, and so $(\alpha,\beta) \in \mathrm{Comp}(G,M)$. Now $(*)$ says that for all $g,h \in G$

$$\beta(\gamma(g,h)) + \kappa(\alpha(gh)) = \delta(\alpha(g),\alpha(h)) + \kappa(\alpha(g))^{\alpha(h)} + \kappa(\alpha(h)).$$

Substituting $g = \alpha^{-1}(u)$ and $h = \alpha^{-1}(v)$ with $u, v \in G$, we deduce that

$$\beta(\gamma(\alpha^{-1}(u),\alpha^{-1}(v))) = \delta(u,v) - \kappa(uv) + \kappa(u)^{\alpha(v)} + \kappa(v)$$

for all $u, v \in G$. This proves $(\alpha,\beta)(\gamma) - \delta = \delta_2(\kappa) \in B^2(G,M)$, as claimed.

Conversely, if $(\alpha,\beta)(\gamma) - \delta = \delta_2(\kappa) \in B^2(G,M)$ for a compatible pair (α,β) and function $\kappa \in C^1(G,M)$, then a short calculation shows that the map $\xi\colon E_\gamma \to E_\delta$, $(g,m) \mapsto (\alpha(g), \beta(m) + \kappa(\alpha(g)))$, defines a strong isomorphism where the required automorphisms of G and M are α and β, respectively. □

Let E_γ be the extension of G by a G-module M defined by $\gamma \in Z^2(G,M)$. Let $\mathrm{Aut}_M(E_\gamma)$ be the subgroup of $\mathrm{Aut}(E_\gamma)$ that contains all automorphisms of E_γ that leave M setwise invariant. Note that $\mathrm{Aut}_M(E_\gamma) = \mathrm{Aut}(E_\gamma)$ if and only if M is characteristic in E_γ.

Every $\xi \in \mathrm{Aut}_M(E_\gamma)$ defines unique $\alpha \in \mathrm{Aut}(G)$ and $\beta \in \mathrm{Aut}(M)$ and a map $\kappa \in C^1(G,M)$ such that ξ maps each $(g,m) \in E_\gamma$ to $(\alpha(g), \beta(m)+\kappa(\alpha(g)))$. Note that this is a strong isomorphism $E_\gamma \to E_\gamma$. The homomorphism property of ξ implies that $(\alpha,\beta) \in \mathrm{Comp}(G,M)$, and therefore $\xi \to (\alpha,\beta)$ defines a homomorphism $\varphi\colon \mathrm{Aut}_M(E_\gamma) \to \mathrm{Comp}(G,M)$. By the first Isomorphism Theorem, $\mathrm{Aut}_M(E_\gamma)$ is an extension of the image of φ by its kernel. The next result discusses both ingredients of this extension.

Proposition 5.14 *Let M be a G-module and $\gamma \in Z^2(G,M)$ a 2-cocycle. Let $\varphi\colon \mathrm{Aut}_M(E_\gamma) \to \mathrm{Comp}(G,M)$, $\xi \mapsto (\xi|_{E_\gamma/M}, \xi|_M)$, be the homomorphism defined above.*

a) *The kernel of φ is abelian and isomorphic $Z^1(G,M)$.*

b) *The image of φ is $\mathrm{Stab}_{\mathrm{Comp}(G,M)}(\gamma + B^2(G,M))$.*

Proof a) Every $\xi \in \ker\varphi$ defines a map $\kappa \in C^1(G,M)$ such that the image of each $(g,m) \in E_\gamma$ under ξ is $(g, m + \kappa(g))$. The homomorphism property of ξ implies that $\kappa \in Z^1(G,M)$. This yields an injective homomorphism $\ker\varphi \to Z^1(G,M)$; it is surjective since every $\kappa \in Z^1(G,M)$ defines an automorphism $E_\gamma \to E_\gamma$, $(g,m) \mapsto (g, m + \kappa(g))$, that lies in $\mathrm{Aut}_M(E_\gamma)$.

b) This is a direct consequence of Proposition 5.13: note that the elements of $\mathrm{Aut}_M(E_\gamma)$ are exactly the strong isomorphisms $E_\gamma \to E_\gamma$. □

5.1.4 Split extensions and complements

An extension $\{0\} \longrightarrow M \xrightarrow{\iota} E \xrightarrow{\pi} G \longrightarrow \{1\}$ *splits* if there is a transversal $\tau\colon G \longrightarrow E$ that is a group homomorphism. In this case the image $K = \tau(G)$ is a complement to $\iota(M)$ in E: this means that $K \leqslant E$ satisfies $E = K\iota(M)$ and $K \cap \iota(M) = \{1\}$. The prototype of a split extension of G by a G-module M is E_0 where $0 \in Z^2(G, M)$ is the 0-cocycle. As in earlier chapters, we denote this by the semidirect product $E_0 = G \ltimes M$. We use $G.M$ to describe an arbitrary extension of G by M.

Lemma 5.15 *Let M be a G-module and $\gamma \in Z^2(G, M)$. The extension E_γ is a split extension of G by M if and only if $\gamma \in B^2(G, M)$.*

Proof If $\gamma \in B^2(G, M)$, then E_γ is equivalent to $E_0 = G \ltimes M$ by Proposition 5.9, hence E_γ is a split extension. Conversely, consider $\gamma \in Z^2(G, M)$ such that E_γ is a split extension. Let $\pi\colon E_\gamma \longrightarrow G$, $(g, m) \mapsto g$, be the natural projection. Let $\tau\colon G \longrightarrow E_\gamma$ be a transversal that is a group homomorphism. Then $\alpha = \pi \circ \tau$ is an automorphism of G, and the transversal $\sigma = \tau \circ \alpha^{-1}$ has the form $\sigma(g) = (g, \kappa(g))$ for some $\kappa \in C^1(G, M)$. Since σ is a group homomorphism, $(g, \kappa(g))(h, \kappa(h)) = (gh, \kappa(gh))$, and so

$$\kappa(gh) = \kappa(g)^h + \kappa(h) + \gamma(g, h)$$

for all $g, h \in G$. This shows that $\gamma = -\delta_2(\kappa) \in B^2(G, M)$, as claimed. □

If E_γ is a split extension, then the set of all complements can be determined explicitly as follows.

Proposition 5.16 *Let M be a G-module and consider a 2-coboundary $\gamma \in B^2(G, M)$. Let $\kappa \in Z^1(G, M)$ be such that $g \mapsto (g, \kappa(g))$ defines a transversal $\tau\colon G \longrightarrow E_\gamma$ that is a group homomorphism, so $K = \tau(G)$ is a complement to M in E_γ. For $\varphi \in Z^1(G, M)$ define $K_\varphi = \{(g, \kappa(g) + \varphi(g)) \mid g \in G\}$.*

a) *The map $\varphi \longrightarrow K_\varphi$ is a bijection between $Z^1(G, M)$ and the set of all complements to M in E_γ.*

b) *Two complements K_φ and K_ϱ are conjugate in E_γ if and only if $\varphi - \varrho$ lies in $B^1(G, M)$.*

Proof The proof of Lemma 5.15 shows that κ and τ exist.

a) Since τ is a group homomorphism, $\kappa(g)^h + \kappa(h) + \gamma(g, h) = \kappa(gh)$ for all $g, h \in G$. A direct calculation now shows that if $\varphi \in Z^1(G, M)$, then K_φ is a subgroup of E_γ. Since τ is a transversal, K_φ is a complement to M in E_γ.

Conversely, let $C \leqslant E_\gamma$ be a complement to M. Now C and $K = \tau(G)$ are both transversals to M in E_γ, so $C = \{(g, \kappa(g) + \varphi(g)) \mid g \in G\}$ for a unique function $\varphi\colon G \longrightarrow M$. Since C is also a group, $\varphi \in Z^1(G, M)$, and so $C = K_\varphi$.

b) Let $\varphi, \varrho \in Z^1(G, M)$ such that $\varphi = \varrho + \delta_1(m)$ for some $m \in C^0(G, M) = M$. Recall that $\delta_1(m)(g) = m - m^g$, so $K_\varphi = \{(g, \kappa(g) + \varrho(g) + m - m^g) \mid g \in G\}$. Since $(g, \kappa(g) + \varrho(g))^{(1,m)} = (g, \kappa(g) + \varrho(g) + m - m^g)$ for every $g \in G$, the element $(1, m)$ conjugates K_ϱ to K_φ.

Conversely, suppose $e \in E_\gamma$ conjugates K_φ to K_ϱ. Since K_φ is a complement to M, we can write $e = km$ with $k \in K_\varphi$ and $m \in M$. Since K_φ is a group, it follows that $K_\varphi^e = K_\varphi^m = K_\varrho$, so $m \in M$ conjugates K_φ to K_ϱ. If $g \in G$, then $(g, \kappa(g) + \varphi(g))^m = (g, \kappa(g) + \varphi(g) + m - m^g)$, and this element is in K_ϱ, hence it must be $(g, \kappa(g) + \varrho(g))$. It follows that $\varrho = \varphi + \delta_1(m)$, as claimed. □

5.2 Sequences in cohomology

Sequences and splittings play an important role in cohomology. Here we recall the results that will be used later.

5.2.1 The long exact cohomology sequence

Let G be a group. A (finite or infinite) sequence

$$\dots \xrightarrow{\alpha_{i-1}} A_i \xrightarrow{\alpha_i} A_{i+1} \xrightarrow{\alpha_{i+2}} \dots .$$

of G-modules A_i and G-module homomorphisms α_i is an *exact sequence* if the image of α_{i-1} equals the kernel of α_i whenever α_{i-1} and α_i exist. The short exact sequence describing a group extension is an exact sequence of $\{1\}$-modules.

The following result provides an important construction of an infinite exact sequence from a short exact sequence.

Proposition 5.17 *Let G be a group and let $\{0\} \to A \xrightarrow{\iota} B \xrightarrow{\pi} C \to \{0\}$ be a short exact sequence of G-modules. There is a long (infinite) exact sequence in cohomology which has the form*

$$\{0\} \longrightarrow H^0(G,A) \xrightarrow{\alpha_0} H^0(G,B) \xrightarrow{\beta_0} H^0(G,C) \xrightarrow{\Lambda_0} H^1(G,A) \xrightarrow{\alpha_1} \dots$$

where each

$$\begin{array}{lll} \alpha_m : H^m(G,A) \to H^m(G,B) & \text{is induced by } \iota, \\ \beta_m : H^m(G,B) \to H^m(G,C) & \text{is induced by } \pi, \text{ and} \\ \Lambda_m : H^m(G,C) \to H^{m+1}(G,A) & \text{is a connecting homomorphism} \end{array}$$

defined as follows: Let $\gamma \in Z^m(G,C)$. There is $\beta \in C^m(G,B)$ such that the map induced by π takes $\beta + B^m(G,B)$ to $\gamma + B^m(G,C)$. The image of $\delta_{m+1}(\beta) \in Z^{m+1}(G,B)$ is contained in A, hence there exists $\alpha \in Z^{m+1}(G,A)$ that maps to $\delta_{m+1}(\beta)$ under the map induced by ι. Now $\Lambda_m(\gamma + Z^m(G,C))$ is defined as $\alpha + Z^{m+1}(G,A)$.

Proof The homomorphisms α_m and β_m have a straightforward explicit definition; the existence of the connecting homomorphisms is a well-known result in homological algebra that is proved using the *Snake Lemma*. Providing details is beyond the scope of this book, so we refer to [44, Theorem IV.2.1] for a proof of the explicit construction; see also [4, Proposition III.6.1]. □

5.2.2 Sequences arising from long cohomology sequences

We need two more sequences; these arise from the *five-term exact sequence* related to a spectral sequence, and the *long exact Ext-sequence*.

Let G be a finite group, let N be a normal subgroup of G, and let A be a G-module. Every $g \in G$ defines a compatible pair $\kappa(g) \in \mathrm{Comp}(N, A)$ where the automorphism on N is conjugation by g and the automorphism on A is the module action of g. Recall from Equation (5.3) that this defines a G-action on $Z^1(N, A)$ and on $H^1(N, A)$. Specifically, if $g \in G$ and $\gamma \in Z^1(N, A)$, then $\kappa(g)(\gamma)\colon N \to A, x \mapsto \gamma(gxg^{-1})^g$. In particular, if $g \in N$, then the cocycle property yields $\gamma(gxg^{-1}) = \gamma(x)+\gamma(g)^x-\gamma(g)$, and γ and $\kappa(g)(\gamma)$ define the same cohomology class in $H^1(N, A)$. This induces a G/N-action on $H^1(N, A)$ where $gN \in G/N$ maps $\gamma + B^1(N, A)$ to $\kappa(g)(\gamma) + B^1(N, A)$.

Proposition 5.18 *Let $N \trianglelefteq G$ and let A be a G-module. Then there is an exact sequence*

$$0 \to H^1(G/N, \mathrm{Fix}_N(A)) \xrightarrow{\alpha} H^1(G, A) \xrightarrow{\beta} \mathrm{Fix}_{G/N}(H^1(N, A)).$$

Proof The maps α and β are the obvious ones: α is induced by the map that takes $\gamma \in Z^1(G/N, \mathrm{Fix}_N(A))$ to the 1-cocycle $G \to A$, $g \mapsto \gamma(gN)$, and β is induced by the map that takes $\gamma \in Z^1(G, A)$ to the 1-cocycle $N \to A$, $g \mapsto \gamma(g)$.

By construction, the image of α lies in the kernel of β. If $\gamma + B^1(G/N, \mathrm{Fix}_N(A))$ is in the kernel of α, then $G \to A$, $g \mapsto \gamma(gN)$, is a 2-coboundary in $B^1(G, A)$, that is, there exists $m \in A$ such that $\gamma(gN) = m - m^g$ for all $g \in G$. For $g \in N$ this implies $0 = m - m^g$, so $m \in \mathrm{Fix}_N(A)$, and therefore γ is a 2-coboundary in $B^1(G/N, \mathrm{Fix}_N(A))$. This shows that α is injective.

Next, we show that the image of β is contained in $\mathrm{Fix}_{G/N}(H^1(N, A))$. Let γ in $Z^1(N, A)$ be the restriction of a 1-cocycle in $Z^1(G, A)$, which, by abuse of notation, we also call γ. Note that $gN \in G/N$ maps γ to $\kappa(g)(\gamma)$, which is the cocycle that maps $x \in N$ to $\gamma(gxg^{-1})^g = \gamma(x) + \gamma(g)^x - \gamma(g)$. This proves that γ and $\kappa(g)(\gamma)$ define the same cohomology class, hence gN fixes $\gamma + B^1(N, A)$ for every $g \in G$.

It remains to show that the kernel of β lies in the image of α. If $\gamma' \in Z^1(G, A)$ defines a cohomology class in the kernel of β, then there exists some $y \in A$ such that $\gamma'(x) = y^x - y$ for all $x \in N$. If $\delta \in B^1(G, A)$ is the 1-coboundary defined by $\delta(g) = -y^g + y$, then $\gamma = \gamma' + \delta \in Z^1(G, A)$ defines the same class as γ', and the restriction of γ to N is 0. Together with the cocycle property of γ, we conclude that for $g \in G$ and $n \in N$

$$\gamma(g) = \gamma(n^{g^{-1}})^g + \gamma(g) = \gamma(n^{g^{-1}} g) = \gamma(gn) = \gamma(g)^n + \gamma(n) = \gamma(g)^n,$$

that is, $\gamma(g) \in \mathrm{Fix}_N(A)$. Hence, γ is induced by a 1-cocycle in $Z^1(G/N, \mathrm{Fix}_N(A))$. This implies that $\ker \beta$ is contained in $\mathrm{im}\, \alpha$. □

We note that the sequence in Proposition 5.18 is part of the five-term exact sequence arising from the spectral sequence corresponding to the group G, normal subgroup N, and module A; we refer to [79, (7.29)] for more details.

Proposition 5.19 *Let P be a finite p-group acting on $T \cong (\mathbb{Z}_p^d, +)$. If $A \leqslant B$ are P-modules, then there is an exact sequence*

$$\mathrm{Hom}_P(T, B) \to \mathrm{Hom}_P(T, B/A) \to H^1(P, \mathrm{Hom}(T, A)).$$

Proof This follows from [44, Propositions IV.7.1 and IV.7.5], together with [57, Theorem 9.4.15]: the series is part of the long exact Ext-sequence in the second variable, associated to the exact sequence $\{0\} \to A \to B \to B/A \to \{0\}$. □

5.3 A cohomology splitting

This section considers the special case of a finite p-group P acting uniserially on the module $T = (\mathbb{Z}_p^d, +)$ with series $T = T_0 > T_1 > \ldots$. A particular aim is to investigate the cohomology groups $H^2(P, T/T_n)$, as these give rise to extensions of P by T/T_n and these extensions, in turn, are constructions of finite p-groups that are of special interest in coclass theory. The results of this section have been introduced first in [32]; we note that they extend to $H^m(P, T/T_n)$ for all $m \geqslant 0$, see Couson [12].

Throughout the section, bounds for orders and exponents of cohomology groups play a significant role. We introduce the following notation:

$$\begin{aligned} v &= \log_p |P|, \\ a &= \log_p \exp(H^2(P, T)), \\ b &= \max\{\log_p \exp(H^3(P, T_i)) \mid i \in \mathbb{N}_0\}, \\ w &= \max\{a, b\}. \end{aligned}$$

Note that b is well-defined, since $T_{i+d} = pT_i$ for $i \in \mathbb{N}_0$ and thus $H^3(P, T_{i+d})$ is isomorphic to $H^3(P, T_i)$. Note that $v \geqslant w$ follows from Proposition 5.2. Bounds for a and b, and therefore also for w, are known. For example, if H is the kernel of the action of P on T, then [32, Lemma 17] shows that p^a divides $p \exp(H/H')$, and p^b divides $p \exp(H/H') \exp(M(H))$, where $M(H)$ is the Schur multiplier of H and H' denotes the derived subgroup.

5.3.1 A splitting of $H^2(P, T/T_n)$

For each $n \in \mathbb{N}_0$, the group T can be considered as an extension of T/T_n by T_n, with corresponding short exact sequence

$$\{0\} \longrightarrow T_n \stackrel{\iota}{\longrightarrow} T \stackrel{\pi}{\longrightarrow} T/T_n \longrightarrow \{0\},$$

where ι is the natural embedding and π is the natural epimorphism with $\pi(t) = t + T_n$ for $t \in T$. In the associated long exact sequence (Proposition 5.17), the following segment is of special interest:

$$\dots \xrightarrow{\alpha_2} H^2(P,T) \xrightarrow{\beta_2} H^2(P,T/T_n) \xrightarrow{\Lambda_2} H^3(P,T_n) \xrightarrow{\alpha_3} H^3(P,T) \xrightarrow{\beta_3} \dots$$

For large enough n, this long exact sequence induces the following short exact sequence.

Proposition 5.20 *Let $n \geqslant dw$. For every $i \in \mathbb{N}$ there is a short exact sequence*

$$\{0\} \xrightarrow{\alpha_i} H^i(P,T) \xrightarrow{\beta_i} H^i(P,T/T_n) \xrightarrow{\Lambda_i} H^{i+1}(P,T_n) \xrightarrow{\alpha_3} \{0\}.$$

Proof Since $T_n \leqslant T_{wd} = p^w T$, every $\gamma \in Z^i(P,T_n)$ can be written as $p^w \gamma'$ for some $\gamma' \in Z^i(P,T)$. Since $H^i(P,T)$ has exponent dividing p^w, it follows that $\alpha_i \colon H^i(P,T_n) \to H^i(P,T)$ is the 0-map. This implies the claim. □

We now consider $H^2(P,T/T_n)$ in more detail and introduce the following filtration for this group. This can also be defined for $H^i(P,T/T_n)$ in general; however, we restrict to the case $i = 2$ since this is all we need.

Definition 5.21 Let $n \in \mathbb{N}$ and $i \in \{0,\dots,n\}$.

a) The group $K_{i,n} \leqslant H^2(P,T/T_n)$ is the image of $H^2(P,T_i/T_n) \to H^2(P,T/T_n)$ which is induced by the inclusion $T_i \to T$, $t \mapsto t$.
b) The group $L_n \leqslant H^2(P,T/T_n)$ is the image of $\beta_2 \colon H^2(P,T) \to H^2(P,T/T_n)$ which is induced by the projection $T \to T/T_n$, $t \mapsto t + T_n$.
c) We write $K_n = K_{\lfloor n/2 \rfloor,n}$.

It is immediate that the groups $K_{i,n}$ define a filtration of $H^2(P,T/T_n)$, that is, a subgroup series

$$H^2(P,T/T_n) = K_{0,n} \geqslant K_{1,n} \geqslant \dots \geqslant K_n \geqslant \dots \geqslant K_{n,n} = \{0\}$$

and K_n is (roughly) the group in the middle of this filtration. The next theorem investigates this further.

Theorem 5.22 *Let $n \geqslant 2dw$.*

a) $K_{dw,n} = K_{dw+1,n} = \dots = K_n = \dots = K_{n-dw,n}$.
b) *The groups L_n and K_n are isomorphic to $H^2(P,T)$ and $H^3(P,T_n)$, respectively.*
c) *The group $H^2(P,T/T_n)$ is the direct sum of L_n and K_n, that is*

$$H^2(P,T/T_n) = L_n \oplus K_n.$$

Proof We prove all three parts simultaneously. Let $i \in \{dw,\dots,n-dw\}$. As $n \geqslant 2dw$, it follows that $n \geqslant dw+i$ and thus $T_n \leqslant T_{i+dw} = p^w T_i$. Hence Proposition 5.20

applies to T and also to T_i. This yields the following commutative diagram of short exact sequences where τ and σ are defined by inclusions.

$$\begin{array}{ccccccccc}
\{0\} & \longrightarrow & H^2(P,T_i) & \xrightarrow{\beta_2'} & H^2(P,T_i/T_n) & \xrightarrow{\Lambda_2'} & H^3(P,T_n) & \longrightarrow & \{0\} \\
 & & \downarrow{\scriptstyle\tau} & & \downarrow{\scriptstyle\sigma} & & \Vert & & \\
\{0\} & \longrightarrow & H^2(P,T) & \xrightarrow{\beta_2} & H^2(P,T/T_n) & \xrightarrow{\Lambda_2} & H^3(P,T_n) & \longrightarrow & \{0\}
\end{array}$$

As $i \geqslant dw$, it follows that $T_i \leqslant p^w T$, and therefore τ is the zero map; see also the argument in the proof of Proposition 5.20. We show that this implies $L_n \cap K_{i,n} = \{0\}$. Recall that $K_{i,n} = \operatorname{im}\sigma$ and $L_n = \operatorname{im}\beta_2$. If $s \in L_n \cap K_{i,n}$, then $s \in \operatorname{im}\beta_2$, hence $s \in \ker\Lambda_2$. Thus, the preimage t of s under σ satisfies $\Lambda_2'(t) = 0$, and therefore $t \in \operatorname{im}\beta_2'$, say $t = \beta_2'(y)$ with $y \in H^2(P,T_i)$. Then $s = \beta_2(\tau(y))$, and since τ is the zero map, $s = 0$, as claimed.

Thus $L_n \oplus K_{i,n} \leqslant H^2(P,T/T_n)$ and it remains to show equality. Consider an element x of $H^2(P,T/T_n)$ lying outside $L_n \oplus K_{i,n}$. Since $L_n = \operatorname{im}\beta_2 = \ker\Lambda_2$, we deduce that $\Lambda_2(x) \neq 0$, and therefore there exists $u \in H^2(P,T_i/T_n)$ with $\Lambda_2'(u) = \Lambda_2(x)$. By construction, x and $\sigma(u)$ map to the same element under Λ_2, so they differ by an element of $\ker\Lambda_2 = L_n$. Since $\sigma(u) \in K_{i,n}$, this implies $x \in L_n \oplus K_{i,n}$, a contradiction. Thus, x cannot exist, so $L_n \oplus K_{i,n} = H^2(P,T/T_n)$.

The claims $K_{i,n} \cong H^3(P,T_n)$ and $L_n \cong H^2(P,T)$ follow from Proposition 5.20 and the first Isomorphism Theorem; note that Λ_2 has kernel L_n and image $H^3(P,T_n)$.

Finally, recall that $K_{dw,n} \geqslant \ldots \geqslant K_{n-dw,n}$. We have shown that each of these groups is a complement to L_n in $H^2(P,T/T_n)$ and hence each is isomorphic to $H^3(P,T_n)$; in particular, they have the same finite order. This proves $K_{dw,n} = \ldots = K_{n-dw,n}$, and so each group equals $K_n = K_{\lfloor n/2\rfloor,n}$. □

Definition 5.23 The splitting of $H^2(P,T/T_n) = L_n \oplus K_n$ obtained in Theorem 5.22 is the *coclass splitting* of this cohomology group.

The coclass splitting induces the following (non-natural) isomorphism of cohomology groups.

Theorem 5.24 *Let $n \geqslant 2dw$.*

a) *The isomorphism $L_n \cong L_{n+d}$ is induced by the identity on $H^2(P,T)$.*

b) *The isomorphism $K_n \cong K_{n+d}$ is induced by multiplication by p.*

c) *Applying the isomorphisms in* a) *and* b) *to the summands of the coclass splitting yields an isomorphism $H^2(P,T/T_n) \to H^2(P,T/T_{n+d})$.*

Proof a) For $k \geqslant n$ let $\beta_{2,k}$ be the embedding $H^2(P,T) \to H^2(P,T/T_k)$ induced by the projection $T \to T/T_k$; recall from Proposition 5.20 that $\beta_{2,k}$ is indeed injective, and by definition $L_n = \operatorname{im}\beta_{2,n}$ and $L_{n+d} = \operatorname{im}\beta_{2,n+d}$.

b) Multiplication by p induces an isomorphism $T_{n-dw}/T_n \to T_{n+d-dw}/T_{n+d}$, which, in turn, induces an isomorphism $\mu_{p,n}\colon K_n \to K_{n+d}$. Alternatively, since $K_n \cong H^3(P,T_n)$, one can also define $K_n \cong K_{n+d}$ as the isomorphism induced by $H^3(P,T_n) \cong H^3(P,T_{n+d})$, which is also defined by multiplication by p.

c) This follows from a), b), and Theorem 5.22c). □

5.3.2 Compatible pairs and the coclass splitting

We investigate the action of $\mathrm{Comp}(P, T/T_n)$ on $H^2(P, T/T_n) = K_n \oplus L_n$ with respect to the coclass splitting.

Recall that a compatible pair (α, β) in $\mathrm{Comp}(P, T)$ satisfies $\beta(t^g) = \beta(t)^{\alpha(g)}$ for all $t \in T$ and $g \in P$. In particular, $\beta([t, g]) = \beta(t^{-1}t^g) = [\beta(t), \alpha(g)]$, which implies that the series $T = T_0 > T_1 > \dots$ is invariant under β. It follows that β induces an automorphism on each quotient T/T_n, denoted $\beta|_{T/T_n}$. For each $n \in N_0$ we obtain the following homomorphism

$$\mathrm{Comp}(P, T) \to \mathrm{Comp}(P, T/T_n), \quad (\alpha, \beta) \mapsto (\alpha, \beta|_{T/T_n})$$

and denote its image by J_n.

Theorem 5.25 *With the previous notation and $n \geqslant 2vd$, the following hold.*

a) *The group K_n is invariant under* $\mathrm{Comp}(P, T/T_n)$.

b) *The group L_n is invariant under J_n.*

Proof a) If $\gamma + B^2(P, T/T_n) \in K_n$, then γ is a 2-cocycle with image in T_{dv}/T_n. If $(\alpha, \beta) \in \mathrm{Comp}(P, T/T_n)$, then the image of $(\alpha, \beta)(\gamma)$ is also contained in T_{dv}/T_n, so $(\alpha, \beta)(K_n) \subseteq K_n$.

b) The group L_n is the image of the homomorphism $H^2(P, T) \to H^2(P, T/T_n)$ induced by the projection $T \to T/T_n$. Since $\mathrm{Comp}(P, T)$ acts on $H^2(P, T)$, the definition of J_n implies that L_n is invariant under the action of J_n. □

For $\alpha \in \mathrm{Aut}(P)$ we denote by T_α the P-module T *twisted* by α; that is, $T_\alpha = T$ as additive abelian group and $g \in P$ acts via $\alpha(g)$ on T. Similarly, we denote by $(T/T_n)_\alpha$ the P-module T/T_n twisted by α. Note that $(T/T_n)_\alpha = T_\alpha/T_n$. This notation permits the following characterisation of $\mathrm{Comp}(P, T/T_n)$:

$$\mathrm{Comp}(P, T/T_n) = \{(\alpha, \beta) \mid \alpha \in \mathrm{Aut}(P),\ \beta \in \mathrm{Hom}_P(T/T_n, (T/T_n)_\alpha) \text{ bijective}\}.$$

The kernel of a P-module homomorphism $T \to (T/T_n)_\alpha$ is P-invariant of index at most $[T : T_n]$, and therefore contains T_n by Proposition 4.4. Thus, there is a natural isomorphism $\mathrm{Hom}_P(T, (T/T_n)_\alpha) \to \mathrm{Hom}_P(T/T_n, (T/T_n)_\alpha)$, which motivates our investigation of $\mathrm{Hom}_P(T, (T/T_n)_\alpha)$.

Theorem 5.26 *Let $n \geqslant 2dv$ and $\alpha \in$* $\mathrm{Aut}(P)$. *Let I_n be the image of the homomorphism* $\mathrm{Hom}_P(T, T_\alpha) \to \mathrm{Hom}_P(T, (T/T_n)_\alpha)$ *induced by the projection $T \to T/T_n$. There exists a subgroup N_n of* $\mathrm{Hom}_P(T, (T_{n-dv}/T_n)_\alpha)$ *such that the following hold.*

a) $\mathrm{Hom}_P(T, (T/T_n)_\alpha) = I_n \oplus N_n$.

b) *Division by p induces an isomorphism $N_{n+d} \to N_n$.*

c) *The projection $T/T_{n+d} \to T/T_n$ induces an epimorphism $I_{n+d} \to I_n$.*

d) *If $\varepsilon \in I_n$ and $\delta \in N_n$, then $\varepsilon + \delta$ is surjective if and only if ε is surjective. The epimorphism in* c) *maps surjective homomorphisms to surjective homomorphisms.*

Proof We prove all parts simultaneously in several steps.

(1). We first determine $\mathrm{Hom}_P(T, T_\alpha)$. We fix a basis of $T = (\mathbb{Z}_p^d, +)$ and identify $\mathrm{Aut}(T)$ with $\mathrm{GL}_d(\mathbb{Z}_p)$ and $\mathrm{End}(T)$ with $\mathbb{Z}_p^{d\times d}$, acting on row vectors. Let $\{g_1, \ldots, g_m\}$ be a generating set of P, and let $\rho_1, \rho_2 \colon P \to \mathrm{GL}_d(\mathbb{Z}_p)$ be the P-actions on the module T and T_α, respectively, such that $\rho_2(g) = \rho_1(\alpha(g))$ for $g \in P$. Then $\mathrm{Hom}_P(T, T_\alpha)$ is the kernel of the $\mathbb{Z}_p$-linear map $\psi = \psi(\rho_1, \rho_2)$ from $\mathrm{End}(T)$ to $\mathbb{Z}_p^{d\times md}$ defined by

$$\psi(M) = \left[\, M\rho_2(g_1) - \rho_1(g_1)M \;\middle|\; \ldots \;\middle|\; M\rho_2(g_m) - \rho_1(g_m)M \,\right].$$

We fix a $\mathbb{Z}_p$-basis of $\mathrm{End}(T) = \mathbb{Z}_p^{d\times d}$ and identify ψ with a $d^2 \times md^2$ matrix over $\mathbb{Z}_p$. Since $\mathbb{Z}_p$ is a principal ideal domain (see Section 3.2), a Smith normal form argument shows that there are invertible $\mathbb{Z}_p$-matrices L and R such that

$$\psi = LDR \quad \text{where} \quad D = \left[\begin{array}{ccc|c} \alpha_1 & & & \\ & \ddots & & 0 \\ & & \alpha_u & \\ \hline & 0 & & 0 \end{array}\right] \in \mathbb{Z}_p^{d^2\times md^2} \tag{5.4}$$

with $\alpha_1, \ldots, \alpha_u \neq 0$ and each $\alpha_i \mid \alpha_{i+1}$. Since the nonzero diagonal entries of D are uniquely defined up to multiplication by elements of $\mathbb{Z}_p^*$, we can assume that each α_i is a prime power, say $\alpha_i = p^{m_i}$ with $m_1 \leqslant \ldots \leqslant m_u$. Note that $\ker D$ is the $\mathbb{Z}_p$-submodule generated by $\{e_{u+1}, \ldots, e_{d^2}\}$ where each e_i is the standard row basis element with 1 in position i. Since R and L are invertible, $\mathrm{Hom}_P(T, T_\alpha) = \ker\psi$ is determined as the $\mathbb{Z}_p$-submodule generated by $\{e_{u+1}L^{-1}, \ldots, e_{d^2}L^{-1}\}$.

(2). We now repeat the computation of Step (1) for $\mathrm{Hom}_P(T, (T/T_n)_\alpha)$. Let V_n be the $\mathbb{Z}_p$-submodule of $\mathbb{Z}_p^{md^2}$ such that $b \in \mathbb{Z}_p^{d^2}$ represents an element of $\mathrm{Hom}_P(T, (T/T_n)_\alpha)$ if and only if

$$(*) \quad b\psi = bLDR \equiv 0 \bmod V_n.$$

We identify the elements of $\mathrm{Hom}_P(T, (T/T_n)_\alpha)$ with such vectors. By construction, $V_{n+d} = pV_n$ for all $n \geqslant d$. It follows from (5.4) that

$$\mathrm{Hom}_P(T, (T/T_n)_\alpha) = I_n \oplus N_n$$

where I_n is the $\mathbb{Z}_p$-submodule generated by $\{e_{u+1}L^{-1}, \ldots, e_{d^2}L^{-1}\}$, and N_n consists of those $\mathbb{Z}_p$-linear combinations b of $\{e_1L^{-1}, \ldots, e_uL^{-1}\}$ with $b\psi \equiv 0 \bmod V_n$. We have seen in (1) that I_n is exactly the subgroup of elements induced by $\mathrm{Hom}_P(T, T_\alpha)$, as claimed. Since both I_n and I_{n+d} are defined as images of $\mathrm{Hom}_P(T, T_\alpha)$, this also proves c). It remains to verify the structure of N_n.

(3). We show that $\exp(N_n)$ divides $|P| = p^\nu$. The projection $T_\alpha \to (T/T_n)_\alpha$ induces $\iota \colon \mathrm{Hom}_P(T, T_\alpha) \to \mathrm{Hom}_P(T, (T/T_n)_\alpha)$ with $\mathrm{im}\,\iota = I_n$. The exact sequence in Proposition 5.19 for $A = (T_n)_\alpha$ and $B = T_\alpha$ now proves that the exponent of $N_n \cong \mathrm{Hom}_P(T, (T/T_n)_\alpha)/\mathrm{im}\,\iota$ divides the exponent of $H^1(P, \mathrm{Hom}(T, (T_n)_\alpha))$, and so it divides $|P| = p^v$ by Proposition 5.2. This implies that image of every homomorphisms in N_n is contained in T_{n-vd}/T_n, which completes the proof of a).

(4). We show that multiplication by p induces a bijection $\mu_p \colon N_n \to N_{n+d}$. Specifically, if $f \in N_n$ maps $t \in T$ to $t' + T_n$, then $\mu_p(f)$ is defined to be the homomorphism $T \to (T_{n-vd+d}/T_{n+d})_\alpha$ that maps t to $pt' + T_{n+d}$. We first confirm that $\mu_p(f)$ lies in N_{n+d}. For this, recall the definition of V_n in Step (2) and let $b \in \mathbb{Z}_p^{d^2}$ be the vector corresponding to f, so that $b\psi \equiv 0 \bmod V_n$ as in $(*)$. Now pb is the vector corresponding to $\mu_p(f)$, and it follows from $pV_n = V_{n+d}$ that $pb\psi \equiv 0 \bmod V_{n+d}$, so $\mu_p(f) \in N_{n+d}$, as claimed. By construction, μ_p is injective; we prove surjectivity in the following two steps.

(4a). We show that $m_1, \ldots, m_u$ (defined in Step (1)) are bounded by v. The matrix $\psi = LDR$ and therefore $m_1, \ldots, m_u$ only depend on P, T, and α, but not on n. To prove that $m_1, \ldots, m_u \leqslant v$, we can therefore choose n large enough such that the integer z with $z \leqslant n/d < z + 1$ satisfies $z > m_1, \ldots, m_u, 2dv$; note that $p^{z+1}T < T_n \leqslant p^z T$. As shown above, a $\mathbb{Z}_p$-linear combination $b = b_1 e_1 L^{-1} + \ldots + b_u e_u L^{-1}$ represents an element in $N_{zd} \leqslant \mathrm{Hom}_P(T, (T/p^z T)_\alpha)$ if and only if $b\psi \equiv 0 \bmod p^z$. This is equivalent to $p^{m_1} b_1 e_1 R + \ldots + p^{m_u} b_u e_u R \equiv 0 \bmod p^z$. Since R is invertible over $\mathbb{Z}_p$ (so every row and column of R contains at least one element of $\mathbb{Z}_p$ not divisible by p), the latter holds if and only if p^{z-m_i} divides b_i for every $i \in \{1, \ldots, u\}$. Now define b to be the element where each $b_i = p^{z-m_i}$, so $b \in N_{zd}$. We saw in Step (3) that N_{zd} has exponent dividing p^v, that is, the homomorphism represented by $p^v b = p^{z-m_1+v} e_1 L^{-1} + \ldots + p^{z-m_u+v} e_u L^{-1}$ induces the zero homomorphism $T \to (T/p^z T)_\alpha$. Since L is invertible over $\mathbb{Z}_p$, the latter holds if and only if p^z divides p^{z-m_i+v} for every i. Thus, $v \geqslant m_1, \ldots, m_u$, as claimed.

(4b). We prove surjectivity of μ_p and consider $b \in N_{n+d}$. If b is represented as $b = b_1 e_1 L^{-1} + \ldots + b_u e_u L^{-1}$, then $b_1 p^{m_1} R + \ldots + b_u p^{m_u} R \equiv 0 \bmod V_{n+d}$. By hypothesis, $n \geqslant 2vd$, which implies that $V_{n+d} \leqslant V_{2vd} = p^{2v} \mathbb{Z}_p^{md^2} \leqslant p^{v+1} \mathbb{Z}_p^{md^2}$. This shows that p^{v+1} divides each $b_i p^{m_i}$. We saw in Step (4a) that $m_1, \ldots, m_u \leqslant v$, and so $p \mid b_1, \ldots, b_u$. This allows us to define $\frac{1}{p}(b_1 e_1 L^{-1} + \ldots + b_u e_u L^{-1})$ as a preimage of μ_p in N_n. Thus, μ_p is surjective, and therefore an isomorphism.

(5). We prove part d). Since the image of δ lies in T_{n-dv}/T_n, it follows that $\varepsilon + \delta$ is surjective if and only if there exists $t \in T$ such that $\varepsilon(t)$ lies in $(T/T_n)_\alpha$ but not in $(T_1/T_n)_\alpha$. By Lemma 4.7, the latter is equivalent to ε being surjective; the claim follows. □

5.4 The coclass of an extension

This section continues to consider the special case that P is a finite p-group of order p^v that acts uniserially on $T = (\mathbb{Z}_p^d, +)$ with series $T = T_0 > T_1 > \ldots$. Our aim is to investigate the coclass of extensions of P by T and of P by T/T_n, respectively.

5.4.1 Extensions of P by T/T_n

For $n \geqslant 2vd$, the coclass splitting Theorem 5.22 proved that $H^2(P,T/T_n) = L_n \oplus K_n$ where $L_n \cong H^2(P,T)$ via the isomorphism β_2, and $K_n \cong H^3(P,T_n)$. The following proposition exploits this.

Proposition 5.27 *Let $n \geqslant 2vd$, let $\lambda \in H^2(P,T)$ and $\gamma \in H^2(P,T/T_n)$ so that there exists $\delta \in K_n$ with $\gamma = \beta_2(\lambda) + \delta$. Then the coclasses of the infinite pro-p-group E_λ and of the finite p-group E_γ coincide; that is; $\mathrm{cc}(E_\lambda) = \mathrm{cc}(E_\gamma)$.*

Proof We consider $T \trianglelefteq E_\lambda$ and $T/T_n \trianglelefteq E_\gamma$ via the usual identifications. By construction and Theorem 5.22, the image of $\delta \in K_n$ is contained in T_{dv}/T_n. This implies that $E_\lambda/T_{dv} \cong E_\gamma/(T_{dv}/T_n)$. Hence $\mathrm{cc}(E_\lambda/T_{dv}) = \mathrm{cc}(E_\gamma/(T_{dv}/T_n))$. It remains to show that $\mathrm{cc}(E_\lambda) = \mathrm{cc}(E_\lambda/T_{dv})$ and $\mathrm{cc}(E_\gamma) = \mathrm{cc}(E_\gamma/(T_{dv}/T_n)$.

First consider E_λ. If $c = \mathrm{cl}(P)$, then $\gamma_{c+1}(E_\lambda) = T_k$ for some $k \leqslant c$ by Proposition 4.4. It follows that $T_{vd} \leqslant T_v < T_c \leqslant T_k = \gamma_{c+1}(E_\lambda)$. The uniserial action of P on T implies that $\mathrm{cc}(E_\lambda/\gamma_{c+1}(E_\lambda)) = \mathrm{cc}(E_\lambda/T_j)$ for all $j \geqslant k$. In particular, $\mathrm{cc}(E_\lambda) = \mathrm{cc}(E_\lambda/T_{vd})$.

The analogous argument shows that $\mathrm{cc}(E_\gamma) = \mathrm{cc}(E_\gamma/(T_{dv}/T_n))$, and so $\mathrm{cc}(E_\lambda) = \mathrm{cc}(E_\lambda/T_{dv}) = \mathrm{cc}(E_\gamma/(T_{dv}/T_n)) = \mathrm{cc}(E_\gamma)$, as claimed. □

If the infinite pro-p-group E_λ in Proposition 5.27 has coclass r, then it defines an infinite path in $\mathcal{G}(p,r)$, see Proposition 1.24, and if $\delta \in K_n$, then $\gamma = \beta_2(\lambda) + \delta$ defines a finite p-group E_γ in $\mathcal{G}(p,r)$. If P is large enough, then E_γ is contained in the maximal coclass tree defined by E_λ. Thus Proposition 5.27 yields a method to construct groups in a maximal coclass tree.

5.4.2 Dimension shifting

Dimension shifting is a well-known process that relates the first and second cohomology group. We adapt this to our setting of finite p-groups. Let P be a finite p-group that acts uniserially on $T = (\mathbb{Z}_p^d,+)$ with series $T = T_0 > T_1 > \ldots$. We write $V = (\mathbb{Q}_p^d,+)$ and extend the series through T to the left by defining $T_{i-d} = p^{-1}T_i$ for $i \in \mathbb{Z}$. Thus, by definition, each T_i is a subgroup of V and $pT_i = T_{i+d}$. We write $F_i = T_{-i}/T$ for $i \in \mathbb{N}_0$.

Proposition 5.28 *Suppose that P acts faithfully on T, and T has dimension $d = d_s$. Let $Z = Z(P)$ be the centre of P of order p^z, write $q = p^{s-z}$ and $F = \mathrm{Fix}_Z(V/T)$.*

a) *The groups $\mathrm{Fix}_Z(T) = \mathrm{Fix}_Z(V) = \{0\}$ are trivial.*

b) *The group F is elementary abelian of rank q, and $F = F_{-q}$.*

c) *The group F is a uniserial P-module with series $F = F_q > \ldots > F_0 = \{0\}$.*

d) *There is an isomorphism $\psi\colon H^2(P,T) \to H^1(P/Z, \mathrm{Fix}_Z(V/T))$.*

Proof The faithful action of P on T allows us to identify P with a subgroup of $\mathrm{GL}_d(\mathbb{Z}_p)$; let I_d be the $d \times d$ identity matrix.

a) Proposition 4.18 shows that Z is cyclic, and V decomposes as a direct sum $V = V_1 \oplus \ldots \oplus V_q$ of q irreducible Z-submodules. A generator g of Z acts diagonally on each V_i as a p^z-th root of unity in dimension $(p-1)p^{z-1}$. Proposition 4.11 asserts that g does not have a fixed point on any V_i.

b) We continue with the argument in the proof of a). As in the proof of Proposition 4.18, the Smith normal form D of $g - I_d$ has q entries p on the diagonal and each other entry is 1. Write $g - I_d = LDR$ with transformation matrices L and R that are invertible over $\mathbb{Z}_p$. The fixed points of g in V/T are those elements $t + T$ with $t \in V$ that satisfy $t^g - t \equiv 0 \bmod T$. This translates to the congruence $(tL)D \equiv 0 \bmod T$. The solutions x to $xD \equiv 0 \bmod T$ form an elementary abelian group of rank q; since L is invertible over $\mathbb{Z}_p$, it follows that F is elementary abelian of rank q.

c) Multiplication by p induces P-module isomorphisms $T_i/T_j \cong T_{i+d}/T_{j+d}$ for all $i \leqslant j$ in $\mathbb{Z}$. Since $F \leqslant V/T$ is finite, there is a p-power p^k such that $p^k F \leqslant T/T_{kd}$. Since $p^k F$ is P-invariant, $p^k F = T_{kd-q}/T_{kd}$ by Proposition 4.2. Thus, $F = T_{-q}/T$, which also shows that P acts uniserially of F with the given series.

d) Since V is uniquely divisible by $|P|$, Proposition 5.2 shows that $H^2(P,V)$ and $H^1(P,V)$ are trivial. Now the long exact sequence (Proposition 5.17) defined by the short exact sequence $\{0\} \to T \to V \to V/T \to \{0\}$ proves that the connecting homomorphism $H^1(P,V/T) \to H^2(P,T)$ is an isomorphism. The same arguments apply to Z, so $H^1(Z,V/T) \cong H^2(Z,T)$. Recall from Proposition 4.18 that Z is cyclic. Proposition 5.2 now implies that $H^2(Z,T)$ is isomorphic to a quotient group of $\mathrm{Fix}_Z(T)$, which is trivial by a). Thus, $H^2(Z,T)$ and $H^1(Z,V/T)$ are trivial. Using this in the exact sequence of Proposition 5.18 (for $N = Z$ and $A = V/T$) yields $H^1(P/Z, \mathrm{Fix}_Z(V/T)) \cong H^1(P,V/T)$, which establishes the claim. □

We explicitly describe the isomorphism in Proposition 5.28d).

Remark 5.29 The isomorphism ψ in Proposition 5.28d) is constructed as the composition of the inverse of the connecting homomorphism $H^1(P,V/T) \to H^2(P,T)$, and the isomorphism $H^1(P,V/T) \to H^1(P/Z, \mathrm{Fix}_Z(V/T)$ induced by $P \to P/Z$. We follow the proof of Proposition 5.28 and explain explicitly how the image $\beta + B^1(P/Z, \mathrm{Fix}_Z(V/T))$ of $\gamma + B^2(P,T)$ under ψ is obtained. Via the embedding $T \to V$, we consider γ as an element of $Z^2(P,V)$. As such, it is a 2-coboundary, say $\gamma = \delta_2(f)$ for some $f\colon P \to V$ with $f(1) = 0$. If $\gamma \notin B^2(P,T)$, then $f \notin Z^1(P,V)$, but f always induces a 1-cocycle $\delta \in Z^1(P,V/T)$ with $\delta(g) = f(g) + T$ for $g \in P$. The proof of Proposition 5.18 shows there exists $\kappa \in B^1(P,V/T)$ such that $\delta' = \delta + \kappa$ satisfies $\delta'(z) = 0$ for $z \in Z$, and now δ' defines $\beta \in Z^1(P/Z, \mathrm{Fix}_Z(V/T))$.

In fact, we can arrange that $\kappa = 0$. If κ is the 1-coboundary defined by $y + T \in V/T$, that is, $\kappa = \delta_1(y + T)$, then we can replace f by $f + \delta_1(y)$, which means we can assume that $\kappa = 0$; recall that $\gamma = \delta_2(f) = \delta_2(f + \delta_1(y))$. In conclusion, the isomorphism ψ maps the class defined by $\gamma \in Z^2(P,T)$ to the class defined by $\beta \in Z^1(P/Z, \mathrm{Fix}_Z(V/T))$ such that

$$\beta(gZ) = \delta'(g) = f(g) + T \quad \text{(for } g \in P\text{)},$$

where $f\colon P \to V$ is a map that satisfies $\delta_2(f) = \gamma$ and $f(z) = 0$ for $z \in Z$.

A corollary of the previous results is that every uniserial p-adic space group can be embedded into a split extension of its point group.

Corollary 5.30 *Let E_γ with $\gamma \in Z^2(P,T)$ be a uniserial p-adic space group with translation subgroup T. There exists $j \in \mathbb{N}_0$ such that the cocycle in $Z^2(P,T_{-j})$ induced by γ defines a split extension $\hat{E}_\gamma$ of P by T_{-j} that contains E_γ as a subgroup.*

Proof Remark 5.29 shows that there exists $f\colon P \to V$ with $\delta_2(f) = \gamma$ and $f(z) = 0$ for $z \in Z$, and the image of f lies in T_{-j} for some $j \in \mathbb{N}$. In particular, γ is a 2-coboundary when considered as a cocycle in $Z^2(P,T_{-j})$, hence the extension E_γ splits when extending the module from T to T_{-j}. Clearly, $E_\gamma \to \hat{E}_\gamma$, $(g,t) \mapsto (g,t)$, is an embedding. □

5.4.3 Extensions of *P* by *T*

We conclude this section with a result that gives an explicit construction for the extensions of P by T with a given coclass. This construction was used in [24] for an algorithm to construct p-adic space groups by coclass. We exhibit the main ideas of the proof here and refer to [57, 58] for details.

Theorem 5.31 *Let P be a group of order p^ν, acting faithfully on T. Let $\gamma \in H^2(P,T)$, let $Z = Z(P)$, and let $F = \mathrm{Fix}_Z(V/T)$ with filtration $F = F_q > \ldots > F_0 = \{0\}$ as in Proposition 5.28. Then there exists $j \in \{0,\ldots,q\}$ with $\mathrm{im}(\psi(\gamma)) \in F_j \setminus F_{j-1}$, and*

$$\mathrm{cc}(E_\gamma) = \nu - j.$$

Proof By Remark 5.29, there is a function $f\colon P \to V$ with $f(z) = 0$ for $z \in Z$, such that $\gamma(g,h) = f(h) - f(gh) + f(g)^h$, and if $\psi(\gamma)$ lies in $H^1(P/Z, F_j)$, but not in $H^1(P/Z, F_{j-1})$, then $f(g) \in T_{-j}$ for all $g \in P$, and j is as small as possible with this property. Recall from Proposition 5.28 that $j \in \{0,\ldots,p^{s-z}\}$ where $|Z| = p^z$. We must show that $\mathrm{cc}(E_\gamma) = \nu - j$, and for this we consider the split supergroup $\hat{E}_\gamma$ of Corollary 5.30. By construction, $\hat{E}_\gamma \cong P \ltimes T_{-j}$. If P has class c and order p^ν, then $\gamma_{c+1}(\hat{E}_\gamma) \cong \gamma_{c+1}(P \ltimes T_{-j}) = T_{-j+c}$, which has index $p^{\nu+c}$ in $P \ltimes T_{-j}$. Since $\hat{E}_\gamma$ acts uniserially on T_{-j}, this implies that $\hat{E}_\gamma$ is an infinite pro-p-group of coclass ν. We use the embedding $E_\gamma \to \hat{E}_\gamma$, $(g,t) \mapsto (g,t)$, to identify E_γ with a subgroup of $\hat{E}_\gamma$ of index p^j. There exists $i \geqslant \max\{j,c\}$ such that $\gamma_{i+1}(E_\gamma) = \gamma_{i+1}(\hat{E}_\gamma) = T_{-j+i}$; this non-trivial result is proved in [58, Proposition 8.7] for odd p, and discussed in [57, p. 242] for $p = 2$. Now $\gamma_{i+1}(E_\gamma) = T_{-j+i}$ implies $\mathrm{cc}(E_\gamma/\gamma_{i+1}(E_\gamma)) = \nu - j$, and since E_γ acts uniserially on $\gamma_{i+1}(E_\gamma)$, it follows that $\mathrm{cc}(E_\gamma) = \nu - j$, as claimed. □

Theorem 5.31 shows how to determine the coclass of an extension of P by T explicitly, but the condition of this theorem is often difficult to evaluate. In the following we exhibit some bounds on the coclass that are straightforward to evaluate; part b) is a famous result of McKay [62].

Theorem 5.32 *Let P be a group of order p^ν and coclass r, acting uniserially on $T \cong (\mathbb{Z}_p^d, +)$. Let S be an extension of P by T.*

a) *The coclass of S satisfies* $\mathrm{cc}(P) \leqslant \mathrm{cc}(S) \leqslant \nu$, *and* $\mathrm{cc}(S) = \nu$ *if* $S = P \ltimes T$.

b) *If P acts faithfully on T, then $d = d_s$ for some $s \in \mathbb{N}$ and $s \leqslant \mathrm{cc}(S)$.*

Proof a) The proof of Proposition 5.27 shows that $\gamma_{c+1}(S) = T_k$ for some $0 \leqslant k \leqslant c$. Thus $\mathrm{cc}(S) = \log_p[S : \gamma_{c+1}(S)] - c = \log_p |P| + k - c = \nu + k - c = \mathrm{cc}(P) + k$. This yields the lower and the upper bound. If S splits, then $k = c$, and so $\mathrm{cc}(S) = \mathrm{cc}(P) + c = \nu$.

b) This is proved in [62]. □

5.5 The isomorphism problem for space groups

Proposition 3.16 introduced the split extension $S(p^\nu) = P \ltimes T$ with $T = (\mathbb{Z}_p[\theta], +)$ and $P = \langle\theta\rangle$ for a primitive p^ν-th root of unity θ over $\mathbb{Q}_p$. We show that this is the unique uniserial p-adic space group with cyclic point group of order p^ν.

Proposition 5.33 *For every $\nu \in \mathbb{N}$, the group $S(p^\nu)$ is, up to isomorphism, the unique uniserial p-adic space group with cyclic point group of order p^ν.*

Proof The existence of $S(p^\nu)$ was established in Proposition 3.16. Now suppose S is a uniserial p-adic space group with cyclic point group P of order p^ν and translation subgroup T of dimension d. It follows from Proposition 5.28 that $\mathrm{Fix}_P(T) = \{0\}$. Proposition 5.2 shows that $H^2(P,T)$ is trivial. Lemma 5.15 shows that S is a split extension, and we can assume that $S = P \ltimes T$. The group $P \leqslant \mathrm{GL}_d(\mathbb{Z}_p)$ is irreducible by Proposition 4.8. The result follows from Proposition 4.11. □

We now consider the isomorphism problem for general extensions that are uniserial p-adic space groups. For this the following notation is required. Let $P \leqslant \mathrm{GL}_d(\mathbb{Z}_p)$ be a finite p-group that acts uniserially on $T = (\mathbb{Z}_p^d, +)$, and denote by $N = N_{\mathrm{GL}_d(\mathbb{Z}_p)}(P)$ the normaliser of P in $\mathrm{GL}_d(\mathbb{Z}_p)$. Recall the definition of the group of compatible pairs $\mathrm{Comp}(P,T)$ from Definition 5.11. A short calculation shows that every $h \in N$ defines a compatible pair (κ_h, μ_h) in $\mathrm{Comp}(P,T)$ where κ_h is conjugation by h and μ_h is the module action of h on T (that is, right multiplication by h); c.f. the discussion preceding Proposition 5.18. Since $\mathrm{Comp}(P,T)$ acts on $H^2(P,T)$, this defines an N-action of $H^2(P,T)$. The next proposition shows that this N-action can be used to solve the isomorphism problem for p-adic space groups with translation subgroup T and point group P.

Proposition 5.34 *With the previous notation, the isomorphism types of uniserial p-adic space groups with translation group T and point group P correspond one-to-one with the* $\mathrm{Comp}(P,T)$*-orbits in $H^2(P,T)$, and with the N-orbits in $H^2(P,T)$*

Proof Proposition 4.32 shows that the translation subgroup of a p-adic space group is characteristic, so the first correspondence follows directly from Proposition 5.13. For the second claim, we prove surjectivity of the map $\varphi\colon N \to \mathrm{Comp}(P, T)$, $h \mapsto (\kappa_h, \mu_h)$, as defined above. Let $(\alpha, \beta) \in \mathrm{Comp}(P, T)$. By definition, if $g \in P$, then $\beta \circ \mu_g = \mu_{\alpha(g)} \circ \beta$. Identifying $\mathrm{Aut}(T)$ with $\mathrm{GL}_d(\mathbb{Z}_p)$, this shows that $\beta = \mu_h$ for a matrix h that normalises P, that is, $h \in N$. Moreover, $\beta \circ \mu_g = \mu_{\alpha(g)} \circ \beta$ implies that $t^{gh} = t^{h\alpha(g)}$ for all $t \in T$ and $g \in P$, that is, $\alpha(g) = h^{-1}gh = \kappa_h(g)$. In conclusion, $(\alpha, \beta) = (\kappa_h, \mu_h)$ for some $h \in N$. □

Chapter 6
The number of infinite paths in $\mathcal{G}(p, r)$

Coclass Theorem D and Proposition 1.24 show that each coclass graph $\mathcal{G}(p, r)$ has only finitely many maximal infinite paths. In this short chapter we investigate *how many* maximal infinite paths exist in $\mathcal{G}(p, r)$; we denote this number by

$$\text{NrPaths}(p, r) = \text{number of maximal infinite paths in } \mathcal{G}(p, r).$$

Exact values are known only in special cases. For coclass 1, there is a unique maximal infinite path for every prime p, so $\text{NrPaths}(p, 1) = 1$; this follows from Proposition 5.33; see also Theorem 9.3. Also, $\text{NrPaths}(2, 2) = 5$ and $\text{NrPaths}(2, 3) = 54$ have been determined by Newman & O'Brien [67], and $\text{NrPaths}(3, 2) = 16$ is shown by Eick et al. [33]. The main result of this chapter is Theorem 6.9, which proves that the following holds for all $r > 1$ and all primes p:

$$\text{NrPaths}(p, r) \leqslant 2r^2 p^{3r(p^{r-1}+3p^{2(r-1)})}.$$

Recall from Proposition 1.24 that $\text{NrPaths}(p, r)$ can also be defined as the number of isomorphism types of infinite pro-p-groups of coclass r. Theorem 4.39 and Corollary 4.40 split the investigation of these groups into two cases: either the group has trivial centre and is a p-adic uniserial space group, or the group has non-trivial centre and is not a space group. Let $\text{NrSpaceGrps}(p, r)$ be the number of isomorphism types of uniserial p-adic space groups of coclass r and let $\text{NrCentExts}(p, r)$ be the number of isomorphism types of infinite pro-p-groups of coclass r with non-trivial centre. Hence,

$$\text{NrPaths}(p, r) = \text{NrSpaceGrps}(p, r) + \text{NrCentExts}(p, r).$$

We investigate $\text{NrSpaceGrps}(p, r)$ and $\text{NrCentExts}(p, r)$ separately, and then combine the results to obtain the desired bound.

The results and methods exhibited in this chapter are based on [23]; we note that the exposition here corrects some minor inconsistencies in [23].

H. Dietrich and B. Eick, *Coclass graphs of p-groups*, Springer Monographs in Mathematics, https://doi.org/10.1007/978-3-032-17506-9_6

6.1 The number of uniserial p-adic space groups

In this section we determine an upper bound for NrSpaceGrps(p,r). Recall from Definition 4.31 and Proposition 4.32 that each uniserial p-adic space group S of coclass r is an extension of a finite point group P by a translation subgroup T; recall that T is characteristic in S and satisfies $T \cong (\mathbb{Z}_p^d, +)$ for the dimension d of S, and $P = S/T$ acts faithfully on T and thus embeds as a uniserial subgroup into $\mathrm{GL}_d(\mathbb{Z}_p)$. Proposition 4.14 and Theorem 5.32 show that $d = d_s = (p-1)p^{s-1}$ for some $s \in \{1, \ldots, r\}$.

Up to conjugacy in $\mathrm{GL}_d(\mathbb{Q}_p)$, the uniserial subgroup P of $\mathrm{GL}_d(\mathbb{Z}_p)$ embeds into W_s or V_s by Proposition 4.14. Our first step in the proof is to determine an upper bound for the number of subgroups of W_s and V_s that can arise for P.

Proposition 6.1 *Let p be a prime and $T = (\mathbb{Z}_p^d, +)$ with $d = d_s$ for some $s \in \mathbb{N}$.*

a) *There are at most $p^{s(1+p+p^2+\ldots+p^{s-1})}$ subgroups in W_s that act uniserially on T.*

b) *There are at most $2^{(s+1)(2^{s-1}+2^{s-3}-1)}$ subgroups in V_s that act uniserially on T for $s \geqslant 3$.*

Proof Recall that $|W_s| = p^{1+p+p^2+\ldots+p^{s-1}}$, and $|V_s| = 2^{2^{s-1}+2^{s-3}-1}$ for $s \geqslant 3$. Propositions 4.26 and 4.30 show that every uniserial subgroup of W_s or V_s can be generated by at most s or $s+1$ elements, respectively. Thus, there are at most $|W_s|^s$ and $|V_s|^{s+1}$ such subgroups in W_s and in V_s, respectively. □

Next, we determine an upper bound to the number of extensions of P by T in the setting of this section.

Proposition 6.2 *Let $P \leqslant \mathrm{GL}_d(\mathbb{Z}_p)$ be a finite p-subgroup that acts uniserially on $T = (\mathbb{Z}_p^d, +)$ where $d = d_s$ for some $s \in \mathbb{N}$. Up to equivalence of extensions, there are at most $p^{sp^{s-1}}$ extensions of P by T for odd p, and at most $p^{(s+1)p^{s-1}}$ extensions for $p = 2$.*

Proof Write $Z = Z(P)$, $V = (\mathbb{Q}_p^d, +)$, and $F = \mathrm{Fix}_Z(V/T)$. Propositions 5.9 and 5.28 prove that the number of equivalence classes of extension of P by T is bounded by $|H^1(P/Z, F)|$. Propositions 4.26 and 4.30 show that P can be generated by s or $s+1$ elements, respectively, depending on p. Since every 1-cocycle in $Z^1(P/Z, F)$ is uniquely determined by its images on a generating set of P/Z, it follows that $|H^1(P/Z, F)| \leqslant |F|^s$ if p is odd and $|H^1(P/Z, F)| \leqslant |F|^{s+1}$ if p is even. Proposition 5.28 shows that $|F|$ divides $p^{(p^{s-1})}$; this implies the claim. □

We can now prove the desired upper bound for NrSpaceGrps(p,r).

Proposition 6.3 *If p is a prime and $r \in \mathbb{N}$, then* $\mathrm{NrSpaceGrps}(p,r) \leqslant 2rp^{3rp^{r-1}}$.

Proof We consider $d = d_s$ for $s \in \{0, \ldots, r\}$. For a fixed d, we estimate the number of $\mathrm{GL}_d(\mathbb{Z}_p)$-classes of uniserial subgroups by considering subgroups P of W_s and V_s that can be generated by s and $s+1$ elements, respectively, and then determine the number of $\mathrm{GL}_d(\mathbb{Z}_p)$-classes in the $\mathrm{GL}_d(\mathbb{Q}_p)$-class of P that have an element in $\mathrm{GL}_d(\mathbb{Z}_p)$. For each such class, we consider all extensions of P by T up to equivalence. The resulting number is an upper bound for the number of isomorphism classes of p-adic uniserial space groups with trivial centre. We make a case distinction.

Odd p. Consider dimension d_s. Proposition 6.1 shows that we need to consider at most $p^{s(1+p+p^2+\ldots+p^{s-1})}$ subgroups of W_s. By Proposition 4.18, each such point group has at most p^{s-1} different lattice actions, and for each of these lattice actions there are at most $p^{sp^{s-1}}$ isomorphism classes of extensions by Proposition 6.2. In summary, this yields

$$\begin{aligned}\mathrm{NrSpaceGrps}(p,r) &\leqslant \sum_{s=1}^{r} p^{s(1+p+p^2+\ldots+p^{s-1})} p^{s-1} p^{sp^{s-1}} \\ &\leqslant rp^{2rp^{r-1}+r(p^{r-1}-1)/(p-1)+(r-1)} \\ &\leqslant rp^{3rp^{r-1}}.\end{aligned}$$

Even p. Consider dimension d_s. The number of space groups whose point group is conjugate to a subgroup of W_s can be bounded as in the odd p case. For the space groups whose point group is conjugate to a subgroup of V_s we use a similar calculation. We need to consider at most $2^{(s+1)(2^{s-1}+2^{s-3}-1)}$ subgroups of V_s by Proposition 6.1, each such group has at most 2^{s-2} lattice actions by Proposition 4.21, and each lattice action gives rise to at most $2^{(s+1)2^{s-1}}$ extensions up to isomorphism. In summary, this yields

$$\begin{aligned}\mathrm{NrSpaceGrps}(2,r) &\leqslant r2^{3r2^{r-1}} + \sum_{s=3}^{r} 2^{(s+1)(2^{s-1}+2^{s-3}-1)} 2^{s-2} 2^{(s+1)2^{s-1}} \\ &\leqslant r2^{3r2^{r-1}} + r2^{(r+1)(2^{r-1}+2^{r-3}-1)} 2^{r-2} 2^{(r+1)2^{r-1}} \\ &\leqslant r2^{3r2^{r-1}} + r2^{2^{r-1}(r+r/4+1+1/4+r+1)} \\ &\leqslant 2r2^{3r2^{r-1}}.\end{aligned}$$

This completes the proof. □

Remark 6.4 The upper bound for $\mathrm{NrSpaceGrps}(p,r)$ obtained in Proposition 6.3 is rather crude. However, the lower bound $p^{p^{r-1}-(r+2)} \leqslant \mathrm{NrSpaceGrps}(p,r)$ was established in [23]. Writing $q = p^{(p^{r-1})}$ and $\ell = p^{r+2}$, we see that

$$\frac{q}{\ell} \leqslant \mathrm{NrSpaceGrps}(p,r) \leqslant 2rq^{3r}.$$

This illustrates that $\mathrm{NrSpaceGrps}(p,r)$ is indeed exponential in p and in r.

6.2 The number of central extensions

Now we consider infinite pro-p-groups G of coclass r that are not space groups. We recall from Theorem 4.39 and Corollary 4.40 that the hypercentre H of G is non-trivial of finite order p^k with $k \leqslant r-1$, and G/H is a uniserial p-adic space group of coclass $r-k$. Moreover, if $N \leqslant Z(G)$ has order p, then G/N is an infinite pro-p-group of coclass $r-1$. Thus, the desired groups G can be constructed as an iterated central extension of pro-p-groups of coclass $r-1, r-2, \ldots, r-k$ by trivial modules. In each iteration step, the extensions correspond to one (or more) elements of a cohomology group $H^2(G, C_p)$. Our first step is to determine an upper bound to the order of this cohomology group.

A *pro-p-polycyclic series* of a pro-p-group H is a subnormal series $H = H_1 \rhd \ldots \rhd H_{n+1} = \{1\}$ of closed subgroups such that each quotient H_i/H_{i+1} is either isomorphic to $(\mathbb{Z}_p, +)$ or to $(C_p, \cdot)$.

Proposition 6.5 *Every infinite pro-p-group of coclass r has a pro-p-polycyclic series of length at most $p^r + p^{r-1} + r$.*

Proof Let G be an infinite pro-p-group of coclass r, so its hypercentre H has order p^k with $0 \leqslant k \leqslant r-1$ and G/H is a uniserial p-adic space group of coclass $r-k$, see Corollary 4.40. The dimension of G/H is $d_s = (p-1)p^{s-1}$ for some $s \in \{1, \ldots, r\}$, see Theorem 5.32. The point group of G/H is isomorphic to a subgroup of W_s or of V_s by Proposition 4.14. Thus, G/H has a pro-p-polycyclic series of length at most $d_s + \max\{\log_p |W_s|, \log_p |V_s|\} = d_s + (1 + p + p^2 + \ldots + p^{s-1})$. Since $|H| = p^k$, the length of a pro-p-polycyclic series of G is bounded from above by $d_s + (1 + p + p^2 + \ldots p^{s-1}) + k \leqslant p^r + p^{r-1} + r$, as claimed. □

The next proposition shows that the length of a pro-p-polycyclic series of a pro-p-group provides an upper bound to the size of the corresponding second cohomology group with coefficients in C_p. Our proof follows established ideas for computing cohomology of (abstract) polycyclic groups; we refer to [46, Section 8.7] for a general discussion.

Proposition 6.6 *Let G be an infinite pro-p-group. If G has a pro-p-polycyclic series of length ℓ, then $|H^2(G, C_p)| \leqslant p^{\ell(\ell+1)/2}$.*

Proof Write $m = \ell(\ell+1)/2$ and let $N = C_p$. Similar to how a polycyclic group with polycyclic series of length ℓ has a power-conjugate presentation with ℓ generators and m relations (see [46, Section 8.1]), a group with pro-p-polycyclic series of length ℓ has a pro-p-presentation with ℓ generators and m relations. Specifically, the group can be described as F/R where F is the free pro-p-group on ℓ generators, and the subgroup R is the topological closure of the normal closure of at most m elements: choose a $\mathbb{Z}_p$-generator for each quotient of the pro-p-polycyclic series and use p-th powers and commutators as relations. We refer to [19, Section 4.6] for more details on pro-p-group presentations.

Suppose G is given by the pro-p-presentation with generators $g_1, \ldots, g_\ell$ and relations $l_i = r_i$ for $i = 1, \ldots, m$; each $l_i = l_i(g_1, \ldots, g_\ell)$ and each $r_i = r_i(g_1, \ldots, g_\ell)$

is a word in the generators with exponents in $\mathbb{Z}_p$. Now consider $\gamma \in Z^2(G,N)$, and denote by $E = E_\gamma$ the associated extension of G by N as described in Definition 5.7. Let $g_1', \ldots, g_\ell' \in E$ be preimages of the generators of G in E, and denote by h a generator of $N \leqslant E$. Since N is central in E, we deduce that $h^p = 1$ and $h^{g_j'} = h$ for $j = 1, \ldots, \ell$, and for every $i \in \{1, \ldots, m\}$ there exists $h_i \in N$ such that $l_i(g_1', \ldots, g_\ell') = r_i(g_1', \ldots, g_\ell')h_i$ in E. This defines a map $\psi\colon Z^2(G,N) \to N^m$, $\gamma \mapsto (h_1, \ldots, h_m)$. Since N is central, ψ is a group homomorphism. A cocycle that lies in the kernel of ψ defines $G \times N$, so $\ker\psi \leqslant B^2(G,N)$ by Lemma 5.15. It follows that $H^2(G,N)$ is isomorphic to $\psi(Z^2(G,N))/\psi(B^2(G,N))$; in particular, $|H^2(G,N)| \leqslant |\psi(Z^2(G,N))| \leqslant p^m$. □

Combining Propositions 6.5 and 6.6 yields an upper bound $e(p,r)$ to $|H^2(G,C_p)|$ for every infinite pro-p-group G of coclass r, that is,

$$|H^2(G,C_p)| \leqslant e(p,r) \quad \text{(independent of } G\text{)}.$$

Lemma 6.7 *Let p be a prime and let $r \in \mathbb{N}$. Then $e(p,r) \leqslant p^{9p^{2r}}$.*

Proof Propositions 6.5 and 6.6 yield $e(p,r) \leqslant p^{\ell(\ell+1)/2}$ for $\ell = p^r + p^{r-1} + r$. Since $\ell(\ell+1)/2 \leqslant \ell^2$ and $\ell \leqslant 3p^r$, the claim follows. □

We now determine an upper bound on the number of central extensions.

Proposition 6.8 *Let p be a prime. Then* $\mathrm{NrCentExts}(p,1) = 0$ *and, if $r > 1$, then*

$$\mathrm{NrCentExts}(p,r) \leqslant \sum_{i=1}^{r-1} \mathrm{NrSpaceGrps}(p,i) \prod_{j=i}^{r-1} e(p,j).$$

Proof Theorem 4.38 shows that every infinite pro-p-group of coclass 1 has trivial centre and thus $\mathrm{NrCentExts}(p,1) = 0$. If $r > 1$, then Theorem 4.39 implies that

$$\begin{aligned}\mathrm{NrCentExts}(p,r) &\leqslant \mathrm{NrPaths}(p,r-1)e(p,r-1)\\ &= (\mathrm{NrSpaceGrps}(p,r-1) + \mathrm{NrCentExts}(p,r-1))e(p,r-1).\end{aligned}$$

The claim follows by induction. □

6.3 An upper bound

The results of the previous sections allow us to prove an upper bound to the number $\mathrm{NrPaths}(p,r)$ of isomorphism types of infinite pro-p-groups of coclass r.

Theorem 6.9 *Let p be a prime. Then* $\mathrm{NrPaths}(p,1) = 1$ *and, if $r > 1$, then*

$$\mathrm{NrPaths}(p,r) \leqslant 2r^2 p^{3r(p^{r-1}+3p^{2(r-1)})}.$$

Proof The claim follows from Propositions 6.3 and 6.8, which imply that

$$\begin{aligned}
\mathrm{NrPaths}(p,r) &= \mathrm{NrSpaceGrps}(p,r) + \mathrm{NrCentExts}(p,r)\\
&\leqslant \sum_{i=1}^{r} 2ip^{3ip^{i-1}} \prod_{j=i}^{r-1} p^{9p^{2j}}\\
&\leqslant \sum_{i=1}^{r} 2rp^{3rp^{r-1}} \prod_{j=1}^{r-1} p^{9p^{2(r-1)}}\\
&= 2r^2 p^{3r(p^{r-1}+3p^{2(r-1)})}.
\end{aligned}$$

Chapter 7
Periodicity I

We established in Theorem 1.27 that each coclass graph $\mathcal{G}(p, r)$ consists of finitely many maximal coclass trees and finitely many groups not contained in these trees. We now investigate the structure of these maximal coclass trees in more detail.

We first introduce *pruned coclass trees*. This idea was used implicitly in Section 2.4 for the graph $\mathcal{G}(5, 1)$. Here we discuss it as a general concept for coclass trees. Then we exhibit the main result of this chapter: a theorem that establishes periodic patterns for pruned coclass trees (*Periodicity I*). These patterns were conjectured by Newman & O'Brien [67] and first proved by du Sautoy [22] using model theory. Eick & Leedham-Green [32] gave an independent second proof using extension theory. A gap in the proof of [32] was discovered and closed by Couson [12]. Our proof follows the ideas of [32] and [12].

We conclude this chapter by exhibiting various applications of Periodicity I. These include a description of *coclass families* and results on finite parametrised presentations for these families, their automorphism groups, their Schur multipliers, and related topics.

7.1 Pruned branches and extension structures

For a maximal coclass tree $\mathcal{T}$ in $\mathcal{G}(p,r)$ let $S_1 \to S_2 \to \dots$ be its mainline, and denote the branches of $\mathcal{T}$ by $\mathcal{B}_1, \mathcal{B}_2, \dots$. Recall that each branch $\mathcal{B}_i$ is a tree with root S_i and it consists of all descendants of S_i in $\mathcal{T}$ that are not descendants of S_{i+1}.

The infinite pro-p-group S associated with the coclass tree $\mathcal{T}$ is the inverse limit of the groups on the mainline of $\mathcal{T}$. Recall that S is an infinite pro-p-group of coclass r and dimension d for some $d \in \mathbb{N}$, see Proposition 1.20. The dimension is an invariant for $\mathcal{T}$ which we also call the dimension of $\mathcal{T}$.

Definition 7.1 The *pruned coclass tree* $\mathcal{T}(k)$ is the subtree of $\mathcal{T}$ generated by all groups of distance at most k from the mainline. The branches in $\mathcal{T}(k)$ are denoted $\mathcal{B}_1(k), \mathcal{B}_2(k), \dots$.

H. Dietrich and B. Eick, *Coclass graphs of p-groups*, Springer Monographs in Mathematics, https://doi.org/10.1007/978-3-032-17506-9_7

As a next step we investigate the extension structure of the groups G in a pruned branch $\mathcal{B}_i(k)$ for fixed $k \in \mathbb{N}$. By construction, such a group G has a normal subgroup N with $G/N \cong S_i$ and $|N| \leqslant p^k$. Recall from Corollary 4.40 that there exists an integer c with $c \leqslant p^{r-1}+1$ such that $S/\gamma_c(S)$ is a mainline group, say $S/\gamma_c(S) \cong S_t$ for some $t \in \mathbb{N}$, and S acts uniserially on $\gamma_c(S) \cong (\mathbb{Z}_p^d, +)$. Based on this notation, we introduce the next result, which is [32, Theorem 15]. The proof uses deep results related to *settled p-groups* and the *degree of commutativity* of finite p-groups. We cite the required results in our proof, and refer to [57, Chapter 11] for a detailed discussion of these concepts.

Theorem 7.2 *Fix $k \in \mathbb{N}$ and $\ell > c + \max\{p^r, 3(k+d)/2\}$. If $G \in \mathcal{B}_i(k)$ for some $i \geqslant \ell$, then $G/\gamma_\ell(G) \cong S/\gamma_\ell(S)$ and the subgroup $\gamma_\ell(G)$ is abelian and isomorphic to $\gamma_\ell(S)/\gamma_{\ell+n}(S)$ as $S/\gamma_\ell(S)$-module, where $|\gamma_\ell(G)| = p^n$.*

Proof Recall that $S/\gamma_c(S) = S_t$, and since $\ell \geqslant c$, the group $S/\gamma_\ell(S) \cong S_{t+\ell-c}$ also lies on the infinite path. Let G be as in the theorem and let $\pi\colon G \to S/\gamma_\ell(S)$ be an epimorphism. Clearly, $\gamma_\ell(G) \leqslant \ker\pi$, and since G and $G/\ker\pi$ both have coclass r, it follows that $\ker\pi = \gamma_\ell(G)$ and G acts uniserially on $\gamma_\ell(G)$. In particular, G is an extension of $S/\gamma_\ell(S)$ by $\gamma_\ell(G)$. The remaining (difficult) part is to prove that the isomorphism between $G/\gamma_\ell(G)$ and $S/\gamma_\ell(S)$ induces a isomorphism between $\gamma_\ell(G)$ and $\gamma_\ell(S)/\gamma_{\ell+n}(S)$ as $S/\gamma_\ell(S)$-modules. Figure 7.1 illustrates the main idea of the proof. In the following we expand the details.

(1). First, we recall the definition of a settled p-group. A finite p-group H is *settled* of dimension d with respect to a normal subgroup N if H acts uniserially on N with series $N = N_1 > N_2 > \ldots > N_{f+1} = \{1\}$, such that $N_i^{[p]} = N_{i+d}$ for all $i \leqslant f - d + 1$, $N_f = \gamma_{\mathrm{cl}(H)}(H)$, and $(p-1)f > pd$. If N is non-abelian, then the corresponding *degree of commutativity* is the largest integer ℓ_H such that $[N_i, N_j] \leqslant N_{i+j+\ell_H}$ for all i, j; otherwise $\ell_H = f - 2$.

(2). We prove that $\gamma_\ell(G)$ is abelian. By assumption, there exists $\ell' \geqslant \ell$ such that G has distance $m \leqslant k$ to $S/\gamma_{\ell'}(S)$. Since $\ell' - c > p^r$, it follows from [57, Theorem 11.2.4] that $S/\gamma_{\ell'}(S)$ is settled with respect to $\gamma_c(S)/\gamma_{\ell'}(S)$. We now define $V_0 \leqslant G$ as the full preimage of $\gamma_c(S)/\gamma_{\ell'}(S)$ under the projection $\pi'\colon G \to S/\gamma_{\ell'}(S)$. The kernel of π' is $\gamma_{\ell'}(G)$ of size p^m with $m \leqslant k$. It is shown in [57, Lemma 11.1.3] that G is settled with respect to V_0, with series

$$V_0 > V_1 > \ldots > V_{\ell'-c+m} = \{1\},$$

such that $\pi'(V_i) = \gamma_{c+i}(S)/\gamma_{\ell'}(S)$ for $i = 0, \ldots, \ell' - c$. The construction also implies that $V_i = \gamma_{c+i}(G)$ for $i = 0, \ldots, \ell' - c + m$; in particular, $V_{\ell-c} = \gamma_\ell(G)$ and $V_{\ell'-c+m} = \{1\}$. By [57, Theorem 11.1.6], the degree of commutativity of G satisfies $\ell_G \geqslant \ell' - c - d$, so $[V_t, V_t] \leqslant V_{2t+\ell'-c-d} = \{1\}$ whenever $t \geqslant (m+d)/2$. Since $k \geqslant m$, it follows that $\gamma_s(G) = V_{s-c}$ is abelian whenever $s \geqslant c + (k+d)/2$. In particular, $\gamma_\ell(G)$ is abelian, as claimed.

(3). Since $\gamma_\ell(G)$ is abelian, the isomorphism $S/\gamma_\ell(S) \cong G/\gamma_\ell(G)$ furnishes $\gamma_\ell(G)$ with the structure of an $S/\gamma_\ell(S)$-module. We show that $\gamma_\ell(G)$ and the quotient $\gamma_\ell(S)/\gamma_{\ell+n}(S)$ are isomorphic as $S/\gamma_\ell(S)$-modules. Let $a \in \mathbb{N}$ satisfy $m \leqslant ad \leqslant m + d$. Since G is settled, $V_i^{[p^a]} = V_{i+ad}$ for $i = 0, \ldots, \ell' - c + m - ad$. Since $ad \leqslant m + d \leqslant k + d$, the assumptions on ℓ imply that $\ell - c - ad \geqslant (k+d)/2$,

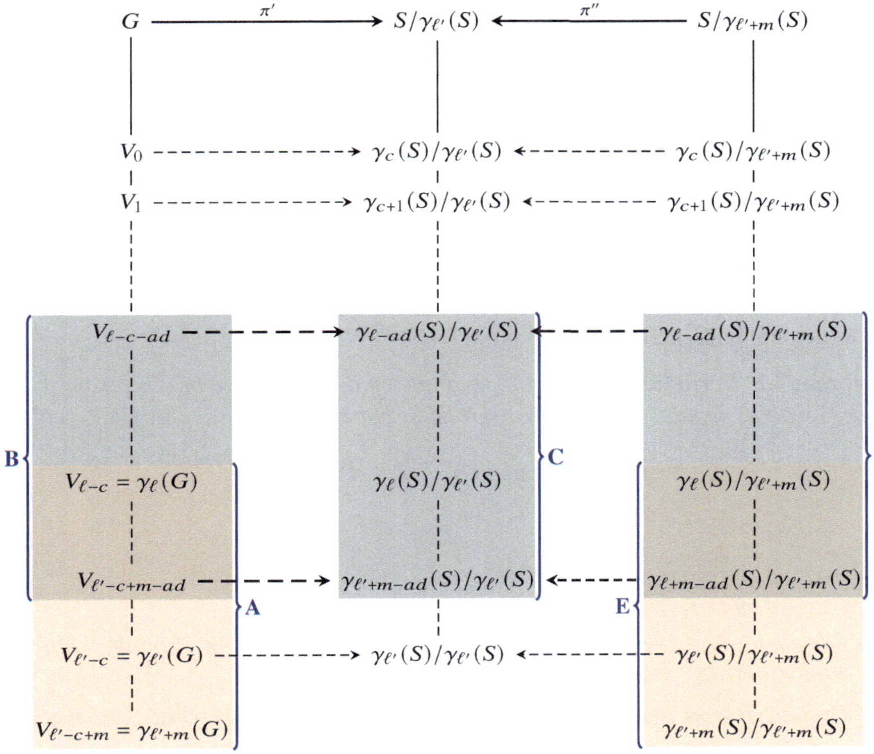

Fig. 7.1 Illustration of the proof of Theorem 7.2

so $V_{\ell-c-ad}$ is abelian by the argument in **(2)**. Thus, exponentiation by p^a yields a $G/\gamma_\ell(G)$-module isomorphism between sections B and A displayed in Figure 7.1, that is,

$$V_{\ell-c-ad}/V_{\ell'-c+m-ad} \cong V_{\ell-c}/V_{\ell'-c+m} \cong V_{\ell-c} = \gamma_\ell(G).$$

Since $\ell' - c + m - ad \leqslant \ell' - c$, the projection π' induces an $S/\gamma_\ell(S)$-module isomorphism between sections B and C in Figure 7.1, that is,

$$V_{\ell-c-ad}/V_{\ell'-c+m-ad} \cong \gamma_{\ell-ad}(S)/\gamma_{\ell'+m-ad}(S),$$

and the projection $\pi''\colon S/\gamma_{\ell'+m}(S) \to S/\gamma_{\ell'}(S)$ induces an isomorphism between sections C and D. Lastly, exponentiation by p^a proves that sections D and E are isomorphic as $S/\gamma_\ell(S)$-modules, that is,

$$\gamma_{\ell-ad}(S)/\gamma_{\ell'+m-ad}(S) \cong \gamma_\ell(S)/\gamma_{\ell'+m}(S).$$

In conclusion, $\gamma_\ell(G)$ and $\gamma_\ell(S)/\gamma_{\ell'+m}(S)$ (with $\ell' + m = \ell + n$) are isomorphic as $S/\gamma_\ell(S)$-modules, as claimed. □

7.2 The theorem of Periodicity I

We now state and prove the main result of this chapter: Periodicity I. It considers a pruned coclass tree $\mathcal{T}(k)$ of dimension d and proves graph isomorphisms

$$\mathcal{B}_i(k) \to \mathcal{B}_{i+d}(k)$$

between its branches, provided that i is sufficiently large depending on $\mathcal{T}$ and k. These isomorphisms are obtained by describing the underlying groups as extensions. This construction, together with explicit bounds on how large i must be, is contained in the following main theorem and its proof.

Theorem 7.3 (Periodicity I). *Let $\mathcal{T}$ be a maximal coclass tree of dimension d in $\mathcal{G}(p,r)$ with associated infinite pro-p-group S. For $k \in \mathbb{N}$ choose $\ell, m, f \in \mathbb{N}$ with*

$$\begin{aligned} \ell &> p^{r-1} + \max\{p^r, 3(k+d)/2\} + 1, \\ m &> 2d(r + \ell - 1), \\ f &> \ell + m. \end{aligned}$$

For $i \in \mathbb{N}_0$ denote $T_i = \gamma_{\ell+i}(S)$ and write $T = T_0$ and $R = S/T$.

a) *For each $i \geqslant \ell$, every group G in $\mathcal{B}_i(k)$ is an extension of R by the natural R-module T/T_n, where $n = \log_p(|G|/|R|)$.*

b) *For each $n \geqslant m$, there is a bijection between the isomorphism types of extensions of R by T/T_n in $\mathcal{T}(k)$ and the isomorphism types of extensions of R by T/T_{n+d} in $\mathcal{T}(k)$.*

c) *For each $i \geqslant f$, the rooted trees $\mathcal{B}_i(k)$ and $\mathcal{B}_{i+d}(k)$ are isomorphic via an isomorphism induced by the bijections in* b).

Proof We first outline the main idea of the proof, and then split the discussion into various steps. Guided by our main aim, part c), we first show in a) that groups in $\mathcal{B}_i(k)$ and in $\mathcal{B}_{i+d}(k)$ are extensions of R by T/T_n and by T/T_{n+d}, respectively, for suitable n. These extensions are described by certain elements of the cohomology groups $H^2(R, T/T_n)$ and $H^2(R, T/T_{n+d})$, and we show that the actions of $\mathrm{Comp}(R, T/T_n)$ and $\mathrm{Comp}(R, T/T_{n+d})$ can be used to decide when two extensions are isomorphic. We use Theorem 5.24 to construct an isomorphism $\xi_{\mathrm{H}}\colon H^2(R, T/T_{n+d}) \to H^2(R, T/T_n)$, and Proposition 5.26 to define an epimorphism $\xi_{\mathrm{C}}\colon \mathrm{Comp}(R, T/T_{n+d}) \to \mathrm{Comp}(R, T/T_n)$. We prove in b) that the isomorphism ξ_{H} induces a bijection between the $\mathrm{Comp}(R, T/T_{n+d})$-orbits in $H^2(R, T/T_{n+d})$ and the $\mathrm{Comp}(R, T/T_n)$-orbits in $H^2(R, T/T_n)$. This bijection will be used to prove c).

a) Corollary 4.40 and our assumptions on ℓ imply that $R = S/\gamma_\ell(S)$ has coclass r; also, S and R act uniserially on $T = \gamma_\ell(S) \cong (\mathbb{Z}_p^d, +)$. In particular, R is a group on the mainline, say $R = S_t$ for some $t \geqslant 1$. This implies $S_j = S/\gamma_{\ell-t+j}(S)$ for $j \geqslant 1$, and $t < \ell$. If G lies in $\mathcal{B}_i(k)$ with $i \geqslant \ell$, then G has a quotient isomorphic to $S_\ell \cong S/\gamma_{2\ell-t}(S)$, and hence also a quotient isomorphic to $S/\gamma_\ell(S) \cong R$. Now a) follows from an application of Theorem 7.2.

b) We split this part of the proof in a sequence of steps.

Step (b1): *Reducing the isomorphism problem to orbits under compatible pairs.* It follows from Proposition 5.8 that every extension G of R by T/T_n in $\mathcal{T}(k)$ is isomorphic to a group extension E_φ for some cohomology class $\varphi \in H^2(R, T/T_n)$; we identify $G = E_\varphi$. Since G and R have the same coclass, $\gamma_\ell(G)$ corresponds to the subgroup T/T_n. As a consequence, every isomorphism between two such extensions E_φ and E_γ in $\mathcal{T}(k)$ is a strong isomorphism that fixes T/T_n. We use Proposition 5.13 to solve the isomorphism problem: two extensions E_φ and E_γ in $\mathcal{B}_i(k)$ are isomorphic if and only if φ and γ lie in the same orbit under $\mathrm{Comp}(R, T/T_n)$.

The analogous analysis applies to extensions of R by T/T_{n+d} in $\mathcal{T}(k)$: these correspond to elements of $H^2(R, T/T_{n+d})$, and the isomorphism problem of such extensions is solved by using the action of $\mathrm{Comp}(R, T/T_{n+d})$.

Step (b2). *Investigating the cohomology groups.* The group R has coclass r and class $\ell-1$. Thus $|R| = p^v$ for $v = r+\ell-1$. The assumptions on m imply that $n \geqslant m > 2dv$. Hence Theorem 5.22 applies and

$$H^2(R, T/T_n) = L_n \oplus K_n,$$

where $L_n \cong H^2(R, T)$ is the image of the embedding $\iota_n\colon H^2(R, T) \to H^2(R, T/T_n)$, and the group $K_n \cong H^3(R, T_n)$ consists of all cohomology classes induced by $Z^2(R, T_{n-dv}/T_n)$. We write elements of L_n as $\iota_n(\varepsilon)$ with $\varepsilon \in H^2(R, T)$. Theorem 5.24 asserts that there is an isomorphism

$$\xi_{\mathrm{H}}\colon H^2(R, T/T_{n+d}) \to H^2(R, T/T_n)$$

that maps $\iota_{n+d}(\varepsilon) \in L_{n+d}$ to $\iota_n(\varepsilon) \in L_n$, and $\delta \in K_{n+d}$ to $\mu_{1/p}(\delta) \in K_n$, where $\mu_{1/p}$ is induced by multiplication by $1/p$.

Step (b3). *Investigating the compatible pairs.* Theorem 5.26 and its preceding discussion show that the compatible pairs in $\mathrm{Comp}(T, T/T_n)$ with first component $\alpha \in \mathrm{Aut}(R)$ have as second component a surjective $\beta \in \mathrm{Hom}_R(T/T_n, (T/T_n)_\alpha)$; recall that $(T/T_n)_\alpha$ denotes the twisted R-module where $g \in R$ acts as $\alpha(g)$. Also, $\mathrm{Hom}_R(T/T_n, (T/T_n)_\alpha)$ is identified with

$$\mathrm{Hom}_R(T, (T/T_n)_\alpha) = I_n \oplus N_n$$

where I_n is the image of $\pi_n\colon \mathrm{Hom}_R(T, T_\alpha) \to \mathrm{Hom}_R(T, (T/T_n)_\alpha)$ and the group N_n contains homomorphisms whose images lie in $(T_{n-dv}/T_n)_\alpha$. We write elements of I_n as $\pi_n(\beta)$ with $\beta \in \mathrm{Hom}_R(T, T_\alpha)$. Theorem 5.26 shows that there is an epimorphism

$$\xi_{\mathrm{C}}\colon \mathrm{Hom}_R(T, (T/T_{n+d})_\alpha) \to \mathrm{Hom}_R(T, (T/T_n)_\alpha)$$

that maps $\pi_{n+d}(\beta) \in I_{n+d}$ to $\pi_n(\beta) \in I_n$, and $\sigma \in N_{n+d}$ to $\mu_{1/p}(\sigma) \in N_n$, where $\mu_{1/p}$ is induced by multiplication with $1/p$. Due to the similar definition, by slight abuse of notation, we denote the latter map by the same name $\mu_{1/p}$ as in Step (b2). We also denote by $\mu^*_{1/p}$ the map that takes $x + T_{n+d}$ for $x \in T_{n+d-dv}$ to

$x/p + T_n \in T_{n-dv}/T_n$, where $|R| = p^v$ as in (b2). Also by abuse of notation, we identify ξ_C with its extension to an epimorphism

$$\xi_C\colon \mathrm{Comp}(R, T/T_{n+d}) \to \mathrm{Comp}(R, T/T_n), (\alpha, \beta) \mapsto (\alpha, \xi_C(\beta)).$$

Step (b4). *Evaluating the action of compatible pairs.* Using the decompositions in Steps (b2) and (b3), we investigate how a compatible pair acts on a cohomology class using the action defined in Equation (5.3). Specifically, let $(\alpha, \pi_n(\beta) + \sigma) \in \mathrm{Comp}(R, T/T_n)$ with $\beta \in \mathrm{Hom}_R(T, T_\alpha)$ bijective and $\sigma \in N_n$; let $\iota_n(\varepsilon) + \delta \in H^2(R, T/T_n)$ with $\varepsilon \in H^2(R, T)$ and $\delta \in K_n$. By linearity, the image of the compatible pair acting on the latter cohomology class can be written as

$$\begin{aligned}&(\alpha, \pi_n(\beta) + \sigma)(\iota_n(\varepsilon) + \delta)\\ &= \underbrace{(\alpha, \pi_n(\beta))(\iota_n(\varepsilon))}_{=\iota_n((\alpha,\beta)(\varepsilon))\in L_n} + \underbrace{(\alpha, \sigma)(\iota_n(\varepsilon))}_{\in K_n} + \underbrace{(\alpha, \pi_n(\beta))(\delta)}_{\in K_n} + \underbrace{(\alpha, \sigma)(\delta)}_{=0}.\end{aligned}$$

We now prove the claimed memberships for the individual summands. To simplify the argument, we consider cocycles instead of cohomology classes

Recall that σ maps T into T_{n-dv}/T_n and δ is induced by a cocycle whose image is contained in T_{n-dv}/T_n. It follows from $T_{n-dv} = p^{-v}T_n$ and $n \geqslant 2dv$ that $T_{n-dv} \leqslant T_{dv} = p^vT$. Thus, $\sigma(T) \subseteq T_{n-dv}/T_n \subseteq p^vT/T_n$, which shows that $\sigma(T_{n-dv}) \subseteq p^vT_{n-dv}/T_n = T_n/T_n$. In particular, we deduce that $(\alpha, \sigma)(\delta) = 0$.

Next, by Theorem 5.25, the subgroup K_n is invariant under $\mathrm{Comp}(R, T/T_n)$, and so $(\alpha, \pi_n(\beta))(\delta) \in K_n$. The definition of σ ensures that $(\alpha, \sigma)(\iota_n(\varepsilon)) \in K_n$.

Lastly, the cocycle $(\alpha, \pi_n(\beta))(\iota_n(\varepsilon))$ maps $g, h \in R$ to $\beta(\varepsilon(\alpha^{-1}(g), \alpha^{-1}(h))) + T_n$, and so $(\alpha, \pi_n(\beta))(\iota_n(\varepsilon)) = \iota_n((\alpha, \beta)(\varepsilon))$ lies in L_n, as claimed.

The corresponding results for $n + d$ are obtained by applying the same reasoning.

Step (b5). *Compatibility of the action with ξ_C and ξ_H.* We now consider a cohomology class $\gamma = \iota_{n+d}(\varepsilon)+\delta$ in $H^2(R, T/T_{n+d})$ and compatible pair $c = (\alpha, \pi_{n+d}(\beta)+\sigma)$ in $\mathrm{Comp}(R, T/T_{n+d})$, and show that

$$\xi_H(c(\gamma)) = \xi_C(c)(\xi_H(\gamma)). \tag{7.1}$$

Since ξ_H is an isomorphism and ξ_C is an epimorphism, Equation (7.1) will imply that ξ_H induces a bijection between the $\mathrm{Comp}(R, T/T_{n+d})$-orbits in $H^2(R, T/T_{n+d})$ and the $\mathrm{Comp}(R, T/T_{n+d})$-orbits in $H^2(R, T/T_{n+d})$. In view of (b1), this will prove b).

It remains to prove Equation (7.1), which we do in the following. Recall that

$$\xi_H(\gamma) = \iota_n(\varepsilon) + \mu_{1/p}(\delta) \quad \text{and} \quad \xi_C(c) = (\alpha, \pi_n(\beta) + \mu_{1/p}(\sigma)).$$

An explicit formula for $C(\gamma)$ was determined in Step (b4), and applying ξ_H expresses the left-hand side of (7.1) as

$$\xi_H(c(\gamma)) = \iota_n((\alpha, \beta)(\varepsilon)) + \mu_{1/p}((\alpha, \sigma)(\iota_{n+d}(\varepsilon))) + \mu_{1/p}((\alpha, \pi_{n+d}(\beta))(\delta)).$$

For the right-hand side of (7.1) we observe

$$\xi_{\mathrm{C}}(c)(\xi_{\mathrm{H}}(\gamma)) = \iota_n((\alpha,\beta)(\varepsilon)) \ + \ (\alpha,\mu_{1/p}(\sigma))(\iota_n(\varepsilon)) \ + \ (\alpha,\pi_n(\beta))(\mu_{1/p}(\delta)).$$

We establish (7.1) by showing that the second and third summands of the previous two equations coincide; we consider cocycles instead of cohomology classes.
(i) We claim $\mu_{1/p}((\alpha,\sigma)(\iota_{n+d}(\varepsilon))) = (\alpha,\mu_{1/p}(v))(\iota_n(\varepsilon))$. Since $\varepsilon \in Z^2(R,T)$, the images of $g, h \in R$ under the two listed cocycles are

$$\mu^*_{1/p}(\ \sigma(\varepsilon(\alpha^{-1}(g),\alpha^{-1}(h)) \ + \ T_{n+d})\ ) \in T_{n-dv}/T_n,$$
$$\mu_{1/p}(\sigma)(\ \varepsilon(\alpha^{-1}(g),\alpha^{-1}(h)) \ + \ T_n \ ) \in T_{n-dv}/T_n.$$

By the definition of $\mu_{1/p}(\sigma)$, these two images coincide, thus (i) holds.
(ii) We claim $\mu_{1/p}((\alpha,\pi_{n+d}(\beta))(\delta)) = (\alpha,\pi_n(\beta))(\mu_{1/p}(\delta))$: Since $\beta\colon T \longrightarrow T_\alpha$, the images of $g, h \in R$ under the two listed cocycles are

$$\mu^*_{1/p}(\ \beta(\delta(\alpha^{-1}(g),\alpha^{-1}(h))) \ + \ T_{n+d} \ ) \in T_{n-dv}/T_n,$$
$$\beta(\ \mu^*_{1/p}(\delta(\alpha^{-1}(g),\alpha^{-1}(h)))\ ) \ + \ T_n \in T_{n-dv}/T_n.$$

Since multiplication by p is an R-module homomorphism, both images coincide, thus (ii) holds. In conclusion, (7.1) holds, so b) is proved.

c) We have seen in b) that for $n \geqslant m$ the isomorphism ξ_{H} induces a bijection between the $\mathrm{Comp}(R,T/T_{n+d})$-orbits in $H^2(R,T/T_{n+d})$ and the $\mathrm{Comp}(R,T/T_n)$-orbits in $H^2(R,T/T_n)$. We now show that for $i \geqslant f$ it also induces a bijection between those orbits of cohomology classes that define groups in the branches $\mathcal{B}_i(k)$ and $\mathcal{B}_{i+d}(k)$, respectively, and, eventually, the graph isomorphism $\mathcal{B}_i(k) \cong \mathcal{B}_{i+d}(k)$. We start by showing that a) and b) are applicable.

Step (c1). *Each group in $\mathcal{B}_i(k)$ satisfies* a) *and* b). Since $i \geqslant f > \ell + m$, part a) applies and every group G in $\mathcal{B}_i(k)$ has R as a quotient. Using the notation of the proof of a), we deduce that $R = S/\gamma_\ell(S) \cong S_t$ with $\ell \geqslant t$, and so $S_j = S/\gamma_{j+\ell-t}(S)$ for all $j \geqslant 1$. This shows that $|S_i|/|R| = p^{i-t}$, and S_i is an extension of R by T/T_n where $n = i - t$. But $i \geqslant \ell + m$ and $\ell \geqslant t$, so $n \geqslant \ell + m - t \geqslant m$. Thus, b) applies to S_i, and therefore also to every other group in $\mathcal{B}_i(k)$ and in $\mathcal{B}_{i+d}(k)$.

Step (c2). *Mainline groups.* Let $n \geqslant m$. We show that if $\varphi = \iota_n(\varepsilon) + \delta \in I_n \oplus K_n$ defines a group in $\mathcal{T}$, then the extension E_ε is isomorphic to our pro-p-group S. Since $\delta \in K_n$, the group $Q = E_\varphi/(T_{n-dv}/T_n)$ is isomorphic to E_ε/T_{n-dv}. Since $Q/\gamma_\ell(Q) \cong R$, we deduce that $\gamma_\ell(E_\varepsilon/T_{n-dv}) = T/T_{n-dv}$, and so $\gamma_\ell(E_\varepsilon) = T = \gamma_\ell(S)$; c.f. Lemma 4.7. Since $E_\varepsilon/\gamma_\ell(E_\varepsilon) \cong R = S/\gamma_\ell(S)$, the group E_ε is an infinite pro-p-group of coclass r. It follows that Q is in the maximal coclass tree defined by both S and E_ε, and so $S \cong E_\varepsilon$ by Proposition 1.24. Since $\gamma_\ell(S) = T = \gamma_\ell(E_\varepsilon)$, every isomorphism $S \longrightarrow E_\varepsilon$ is a strong isomorphism. This shows that ε is unique up to the action of $\mathrm{Comp}(R,T)$ on $H^2(R,T)$, see Proposition 5.13.

Step (c3). *The isomorphisms ξ_{H} preserve edges.* For each $s > m$, the projection $T/T_s \to T/T_{s-1}$ induces a homomorphism $\tau_s \colon H^2(R, T/T_s) \to H^2(R, T/T_{s-1})$. We denote by $\xi_{\mathrm{H},s}$ our isomorphism ξ_{H} from $H^2(R, T/T_{s+d})$ to $H^2(R, T/T_s)$. If $\varphi \in H^2(R, T/T_s)$ defines a group E_φ in our coclass tree $\mathcal{T}$, then $\gamma(E_\varphi)$ corresponds to the subgroup T_{s-1}/T_s, and so the parent (immediate ancestor) of E_φ is the group defined by $\tau_s(\varphi) \in H^2(R, T/T_{s-1})$. By construction, the homomorphisms τ_s are compatible with our maps $\xi_{\mathrm{H},s}$, in the sense that $\tau_s \circ \xi_{\mathrm{H},s+d} = \xi_{\mathrm{H},s+d-1} \circ \tau_{s+d}$.

Step (c4). *The isomorphisms ξ_{H} preserve branches.* Consider a group at depth $j \geqslant 0$ in the branch $\mathcal{B}_{i+d}(k)$, say E_φ for some $\varphi = \iota_{n+d}(\varepsilon) + \delta$ in $H^2(R, T/T_{n+d})$ with $\delta \in K_{n+d}$. Write $\varphi' = \xi_{\mathrm{H}}(\varphi)$ and recall that $\varphi' = \iota_n(\varepsilon) + \mu_{1/p}(\delta)$. By b), both φ and φ' define groups in $\mathcal{T}(k)$, and, by (c2), both $\iota_n(\varepsilon)$ and $\iota_{n+d}(\varepsilon)$ define mainline groups. If $j = 0$, then $E_\varphi \cong S_{i+d}$ is the root of the branch, and by Step (c2) we can assume that $\varphi = \iota_{n+d}(\varepsilon)$. Then the extension defined by $\varphi' = \iota_n(\varepsilon)$ is the root S_i of $\mathcal{B}_i(k)$. Now consider $j \geqslant 1$. Note that S_{i+d} is the largest mainline quotient of E_φ. By construction, $E_{\varphi'}$ has S_i as a quotient, and we now show that it does not have S_{i+1} as a quotient: this will prove that $E_{\varphi'}$ has depth j in the branch $\mathcal{B}_i(k)$, as claimed. By Step (c3), it is sufficient to consider $j = 1$. Suppose, for a contradiction, that E_φ has depth 1 in $\mathcal{B}_{n+d}(k)$, but $E_{\varphi'}$ lies outside $\mathcal{B}_i(k)$, that is, $E_{\varphi'} \cong S_{i+1} \cong E_{\iota_n(\varepsilon)}$. In this case, φ' and $\iota_n(\varepsilon)$ define isomorphic extensions, hence they are in the same $\mathrm{Comp}(R, T/T_n)$-orbit. Now we deduce from Step (b5) that φ and $\iota_{n+d}(\varepsilon)$ are in the same $\mathrm{Comp}(R, T/T_{n+d})$-orbit, hence they define isomorphic extensions $E_\varphi \cong E_{\iota_{n+d}(\varepsilon)} \cong S_{n+d+1}$. This contradicts our assumption that $E_\varphi \in \mathcal{B}_{i+d}(k)$, and therefore $E_{\varphi'}$ has depth 1 in $\mathcal{B}_i$, as claimed.
In conclusion, ξ_{H} defines a bijection between the groups at depth j in $\mathcal{B}_{i+d}(k)$ and the groups at depth j in $\mathcal{B}_i(k)$. Together with (c3), these bijections induce the graph isomorphism $\mathcal{B}_{i+d} \cong \mathcal{B}_i(k)$. This proves part c). □

Theorem 7.3 exhibits explicit bounds for Periodicity I in coclass graphs. We now compare these bounds with experimental observations. Recall that each of $\mathcal{G}(2,1)$ and $\mathcal{G}(3,1)$ has a single maximal coclass tree whose vertices have distance at most $k = 1$ from the mainline, see Figures 1.5 and 1.6. Using the notation of Theorem 7.3, Figure 7.2 exhibits the lower bounds for ℓ, m, f and contrasts these with the start i of the experimentally observed periodicities $\mathcal{B}_i \cong \mathcal{B}_{i+d}$.

graph	k	$\ell >$	$m >$	$f >$	start of observed periodicity
$\mathcal{G}(2,1)$	1	6	13	20	2
$\mathcal{G}(3,1)$	1	7	29	37	3

Fig. 7.2 Lower bounds for the periodicity in the maximal coclass trees of $\mathcal{G}(2,1)$ and $\mathcal{G}(3,1)$

The graph $\mathcal{G}(2,2)$ has maximal coclass trees $\mathcal{T}_1, \ldots, \mathcal{T}_5$, and 16 other groups, see Figure 2.2. We list the lower bounds for these trees in Figure 7.3.

graph	k	d	$\ell >$	$m >$	$f >$	start of observed periodicity
$\mathcal{T}_1$	2	2	10	48	59	1
$\mathcal{T}_2$	2	2	10	48	59	1
$\mathcal{T}_3$	1	1	8	19	28	1
$\mathcal{T}_4$	2	1	8	19	28	1
$\mathcal{T}_5$	1	1	8	19	28	2

Fig. 7.3 Lower bounds for the periodicity in the maximal coclass trees of $\mathcal{G}(2, 2)$

7.3 Coclass families and parametrised presentations

Theorem 7.3 shows that for each maximal coclass tree $\mathcal{T}$ of dimension d with associated infinite pro-p-group S and each fixed $k \in \mathbb{N}$ there exists an integer $f = f(p, r)$ such that there are graph isomorphisms for each $i \geqslant f$:

$$\sigma_i : \mathcal{B}_i(k) \to \mathcal{B}_{i+d}(k).$$

Thus, there are at most d isomorphism types of pruned branches $\mathcal{B}_i(k)$, $i \geqslant f$, with representatives among $\mathcal{B}_f(k), \ldots, \mathcal{B}_{f+d-1}(k)$. Note that there may be several graph isomorphisms $\mathcal{B}_i(k) \to \mathcal{B}_{i+d}(k)$. We specify σ_i to be the graph isomorphism constructed in the proof of Theorem 7.3: each group G in $\mathcal{B}_i(k)$ arises as an extension of the mainline group $R = S/\gamma_\ell(S)$ by the module T/T_n, where $T = \gamma_\ell(S)$ and $T_n = \gamma_{\ell+n}(S)$ and $n = \log_p(|G|/|R|)$. This extension is defined by a cohomology class in $H^2(R, T/T_n)$. The graph isomorphisms σ_i are induced by the inverses of the isomorphisms of Theorem 5.24; these inverses have the form

$$\xi_n : H^2(R, T/T_n) \to H^2(R, T/T_{n+d}).$$

Note that the subscripts in σ_i and ξ_n are clear from the context and thus, to shorten notation, we write σ and ξ for the respective maps.

Definition 7.4 Let $G \in \mathcal{B}_i(k)$ for some $f \leqslant i < f + d$. Define $G_0 = G$ and, iteratively, $G_{s+1} = \sigma(G_s) \in \mathcal{B}_{i+sd}(k)$ for $s \in \mathbb{N}$. The sequence $(G_s \mid s \in \mathbb{N}_0)$ is the *coclass family* defined by G; see Figure 7.4 for an illustration.

It follows that, with the finitely many exceptions of the groups in $\mathcal{B}_i(k)$ for $i < f$, all groups in the pruned tree $\mathcal{T}(k)$ are partitioned into finitely many coclass families corresponding one-to-one to the groups in the pruned branches $\mathcal{B}_f(k), \ldots, \mathcal{B}_{f+d-1}(k)$. Thus we have proved the following.

Proposition 7.5 *Each pruned maximal coclass tree $\mathcal{T}(k)$ in $\mathcal{G}(p, r)$ consists of finitely many infinite coclass families and finitely many other groups.*

As an example, the (unique) maximal coclass tree in $\mathcal{G}(2, 1)$ has three coclass families: the family of dihedral groups, the family of quaternion groups, and the family of semidihedral groups. We saw in Equation (1.1) that the groups in each of these infinite families can be described by a single parametrised presentation. The following theorem shows that this feature holds in general.

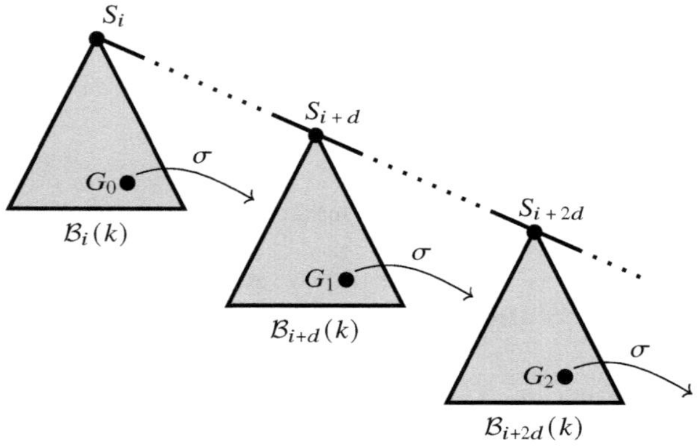

Fig. 7.4 Illustrating Definition 7.4: the coclass family $(G_s \mid s \in \mathbb{N}_0)$ in a maximal coclass tree.

Theorem 7.6 *Let p be a prime and let $r, k \in \mathbb{N}$. The infinitely many groups in a coclass family $(G_s \mid s \in \mathbb{N})$ of a pruned maximal coclass tree $\mathcal{T}(k)$ in $\mathcal{G}(p, r)$ can be described by a single parametrised pro-p-presentation with one integer parameter.*

Proof Let S be the infinite pro-p-group associated with $\mathcal{T}$ and choose ℓ as in Definition 7.4. Let $T_i = \gamma_{\ell+i}(S)$ for $i \in \mathbb{N}_0$ and $T = T_0$. Write $R = S/T$. Then $G = G_0$ is an extension of R with T/T_n via $\varphi \in H^2(R, T/T_n)$. The other groups in the coclass family arise as extensions in a similar style: G_s is an extension of R by T/T_{n+sd} via the cohomology class that arises by s-fold application of ξ to φ.

Recall that $H^2(R, T/T_{n+sd}) = L_{n+sd} \oplus K_{n+sd}$ by Theorem 5.22. Thus $\varphi = \iota_n(\varepsilon) + \delta$ with $\varepsilon \in H^2(R, T)$ and $\delta \in K_n$ and $\xi^s(\varphi) = \iota_{n+sd}(\varepsilon) + \mu_{p^s}(\delta)$, where μ_{p^s} is induced by multiplication with p^s.

The cohomology class ε defines the infinite pro-p-group S as an extension of R by T. A pro-p-presentation for S exhibiting this extension is obtained as follows. Let $R = \langle g_1, \ldots, g_u \mid r_1, \ldots, r_v \rangle$. Let T be generated by $\{t_1, \ldots, t_d\}$ as $\mathbb{Z}_p$-module, written multiplicatively. The action of R on the R-module T can be described by writing $t_i^{g_j}$ as a word in $t_1, \ldots, t_d$ for $1 \leqslant i \leqslant d$ and $1 \leqslant j \leqslant u$; that is, $t_i^{g_j} = m_{i,j} = m_{i,j}(t_1, \ldots, t_d)$. We note that the word $m_{i,j}$ has the form $t_1^{e_1} \cdots t_d^{e_d}$ with $e_i \in \mathbb{Z}_p$. Further, there is a natural epimorphism $S \to R$, and we choose a preimage h_i of each g_i in S. We evaluate each relator $r_1, \ldots, r_v$ of R in these preimages and thus obtain $x_1, \ldots, x_v \in T$. We write each x_i as a (multiplicative) word u_i in $t_1, \ldots, t_d$. This yields a pro-p-presentation (cf. Proposition 6.6) for S on the generators $g_1, \ldots, g_u, t_1, \ldots, t_d$ with relators

$$
\begin{aligned}
& r_i/u_i && \text{for } i \in \{1, \ldots, v\}, \\
& t_i^{g_j}/m_{i,j} && \text{for } i \in \{1, \ldots, d\} \text{ and } j \in \{1, \ldots, u\}, \\
& [t_i, t_j] && \text{for } 1 \leqslant i < j \leqslant d.
\end{aligned}
$$

A similar process can be applied to determine presentations for the extensions G_s of R by T/T_{n+sd} via $\xi^s(\varphi)$. For this, let $w_1, \ldots, w_d \in T$ be a $\mathbb{Z}_p$-generating set for T_n and write each $w_i = w_i(t_1, \ldots, t_d)$ as a word in $t_1, \ldots, t_d$ with exponents in $\mathbb{Z}_p$. Then $\{w_i^{(p^s)} \mid 1 \leqslant i \leqslant d\}$ is a $\mathbb{Z}_p$-generating set for T_{n+sd}. Note that each $w_i^{(p^s)}$ is a word in $t_1, \ldots, t_d$ with exponents in $p^s\mathbb{Z}_p$. Next, the cohomology class δ defines $y_1, \ldots, y_v \in T$ such that for each $s \in \mathbb{N}_0$ the group G_s has a pro-p-presentation on the generators $g_1, \ldots, g_u, t_1, \ldots, t_d$ with relations of the form

$$\begin{array}{ll} r_i/(x_i y_i^{(p^s)}) & \text{for } i \in \{1, \ldots, v\}, \\ t_i^{g_j}/m_{i,j} & \text{for } i \in \{1, \ldots, d\} \text{ and } j \in \{1, \ldots, u\}, \\ [t_i, t_j] & \text{for } 1 \leqslant i < j \leqslant d, \\ w_i^{(p^s)} & \text{for } i \in \{1, \ldots, d\}. \end{array}$$

By construction, each of x_i, y_i, $m_{i,j}$, w_i is an element of T, that is, a word in $t_1, \ldots, t_d$ with exponents in $\mathbb{Z}_p$. However, for $s \in \mathbb{N}_0$, the group G_s is finite and these elements in T can be rewritten, modulo T_{n+sd}, as words having non-negative integer coefficients. Once this rewriting is executed, we obtain a finite presentation with single parameter s for G_s. □

We note that the presentations resulting from Theorem 7.6 may be proper pro-p-presentations. Consider the space group from Proposition 4.12: no integral description exists for the action of Q_{16} on $(\mathbb{Z}_2^4, +)$.

Example 7.7 We give one application of Theorem 7.6. Consider the infinite pro-3-group $S = C_3 \ltimes (\mathbb{Z}_3^2, +)$ of coclass 1 and dimension $d = 2$ with pro-3-presentation

$$S = \langle\, g, z_1, z_2 \mid g^3,\ z_1^g = z_2,\ z_2^g = z_1^{-1} z_2^{-1},\ z_2^{z_1} = z_2 \,\rangle_{\mathbb{Z}_3}.$$

Set $R = S/\gamma_3(S)$ and $T = \gamma_3(S)$. Since $\gamma_3(S) = \langle z_1^3, z_2^3 \rangle_{\mathbb{Z}_3}$, we can define R as

$$R = \langle\, g_1, g_2, g_3 \mid g_1^3,\ g_2^3,\ g_3^3,\ g_2^{g_1} = g_3,\ g_3^{g_1} = g_2^2 g_3^2\,,g_3^{g_2} = g_3 \rangle$$

and consider $T = \langle t_1, t_2 \rangle_{\mathbb{Z}_3}$ where $t_1 = z_1^3$ and $t_2 = z_2^3$. The R-module action on R is defined by $t_1^{g_1} = t_2$, $t_2^{g_1} = t_1^{-1} t_2^{-1}$, and $t_i^{g_j} = t_i$ for $i \in \{1,2\}$ and $j \in \{2,3\}$. Note that S is isomorphic to the group given by the pro-3-presentation with generators g_1, g_2, g_3, t_1, t_2 and relations

$$g_1^3 = x_1,\ \ g_2^3 = x_2,\ \ g_3^3 = x_3,\ \ g_2^{g_1} = g_3 x_4,\ \ g_3^{g_1} = g_2^2 g_3^2 x_5,\ \ g_3^{g_2} = g_3 x_6,$$
$$t_2^{t_1} = t_2,\ \ t_1^{g_1} = t_2,\ \ t_2^{g_1} = t_1^{-1} t_2^{-1},\ \ t_i^{g_j} = t_i\ \ (i \in \{1,2\}, j \in \{2,3\})$$

where $x_1 = x_4 = x_6 = 1$, $x_2 = t_1$, $x_3 = t_2$, and $x_5 = t_1^{-1} t_2^{-1}$. Since $T_3 = \langle t_1^3 t_2^6, t_2^9 \rangle_{\mathbb{Z}_3}$, for every $s \in \mathbb{N}_0$

$$T_{3+2s} = \langle t_1^{3\cdot 3^s} t_2^{6\cdot 3^s}, t_2^{9\cdot 3^s} \rangle_{\mathbb{Z}_3}.$$

By construction, T corresponds to $\gamma_3(S)$, so the mainline group $S/\gamma_{6+2s}(S)$ is an extension of R by T/T_{3+2s}, which is defined by the above presentation with added relators $(t_1^3 t_2^6)^{3^s}$ and $(t_2^9)^{3^s}$. In this group, we can now also rewrite $t_1^{-1}t_2^{-1}$ as $t_1^{3\cdot 3^s-1}t_2^{6\cdot 3^s-1}$. In conclusion, the mainline quotients $S/\gamma_{6+2s}(S)$ with $s \in \mathbb{N}_0$ can be described by the parametrised group presentation with abstract generators g_1, g_2, g_3, t_1, t_2 and the following defining relations:

$$g_1^3 = 1,\ \ g_2^3 = t_1,\ \ g_3^3 = t_2,\ \ g_2^{g_1} = g_3,\ \ g_3^{g_1} = g_2^2 g_3^2 t_1^{3\cdot 3^s-1} t_2^{6\cdot 3^s-1},\ \ g_3^{g_2} = g_3$$

$$t_2^{t_1} = t_2,\ \ t_1^{g_1} = t_2,\ \ t_2^{g_1} = t_1^{3\cdot 3^s-1} t_2^{6\cdot 3^s-1},\ \ t_i^{g_j} = t_i \quad (i \in \{1,2\}, j \in \{2,3\}),$$

$$t_1^{3\cdot 3^s} = t_2^{3\cdot 3^s},\ \ t_2^{9\cdot 3^s} = 1.$$

The resulting presentation is a polycyclic presentation with parameter $s \in \mathbb{N}_0$.

Theorems 7.3 and 7.6 represent a remarkable milestone. While it seems highly unlikely that p-groups can be classified by order, it is possible to achieve a classification of the groups in a pruned maximal coclass tree. Additionally, this classification shows that these groups have remarkable similarities. We conclude this chapter with a discussion of further commonalities among the infinitely many groups within a coclass family.

7.4 Automorphism groups and coclass families

We consider the automorphism groups of the groups in a coclass family $(G_s \mid s \in \mathbb{N}_0)$ in a pruned maximal coclass tree $\mathcal{T}(k)$ of dimension d. By construction, $|G_{s+1}| = p^d|G_s|$ for each $s \in \mathbb{N}_0$. Couson [10,11] and Eick [25] showed that a similar result holds for the respective automorphism groups, see Proposition 7.8. Recall that G_s is an extension of a finite p-group R by a module T/T_{n+sd}, and the subgroup of G_s corresponding to T/T_{n+sd} is characteristic in G_s. Due to this structure, we can use Proposition 5.14 to investigate the corresponding automorphism groups. This requires the consideration of compatible pairs and their action on cohomology groups, as defined in Equation (5.3). Let $e \in \mathbb{N}_0$ be defined by $|\mathrm{Comp}(R, T/T_{n+d})| = p^e|\mathrm{Comp}(R, T/T_n)|$.

Proposition 7.8 *For every $s \in \mathbb{N}_0$ the groups in the family $(G_s \mid s \in \mathbb{N}_0)$ satisfy $|\mathrm{Aut}(G_{s+1})| = p^{d+e}|\mathrm{Aut}(G_s)|$.*

Proof As in the proof of Theorem 7.6, we assume that each group G_s is defined by the cocycle $\xi^s(\varphi) = \iota_{n+sd}(\varepsilon) + \mu_{p^s}(\delta)$ for some $\varepsilon \in Z^2(R,T)$ defining our pro-p-group S and $\delta \in Z^2(R, T/T_n)$ with image in T_{n-dv}/T_n where $|R| = p^v$. By construction, $\gamma_\ell(G_s)$ corresponds to the subgroup T/T_{n+sd} of the extension, which shows that every automorphism $G_s \to G_s$ is a strong isomorphism with respect to the normal subgroup T/T_{n+sd}. This allows us to apply Proposition 5.14 to determine $\mathrm{Aut}(G_s)$ by considering the homomorphism $\psi_s\colon \mathrm{Aut}(G_s) \to \mathrm{Aut}(R) \times \mathrm{Aut}(T/T_{n+sd})$: the image of ψ_s is the stabiliser in $\mathrm{Comp}(R, T/T_{n+sd})$ of the cohomology class of $\xi^s(\varphi)$ and the kernel of ψ_s is isomorphic to $Z^1(R, T/T_{n+sd})$. To simplify the notation, in the following we write

$$K_s = \ker(\psi_s), \quad J_s = \mathrm{im}(\psi_s), \quad \text{and} \quad M_s = T/T_{n+sd}.$$

First, we show that $|K_{s+1}| = p^d|K_s|$ for $s \in \mathbb{N}_0$. By Proposition 5.14, the order of K_s is $|Z^1(R, M_s)| = |H^1(R, M_s)||B^1(R, M_s)|$. Since $n \geqslant 2dv$, Proposition 5.20 shows that $|H^1(R, M_s)| = |H^1(R, T)||H^2(R, T_{n+sd})|$, which is independent of s since $T_{n+sd} = p^s T_n$. (This follows also from the proof of Theorem 5.22 adapted to $H^1(R, M_s)$.) The group $B^1(R, M_s)$ is the image of the map $\tau\colon M_s \to C^1(R, M_s)$, $m \mapsto \delta_m$, where δ_m is the map that takes g to $m^g m^{-1}$. The kernel of τ consists of the fixed points in M_s under R. Since M_s is a finite uniserial R-module, $[R, \ker(\tau)] = \{1\}$ and so $|\ker(\tau)| = p$. By the first Isomorphism Theorem, $|B^1(R, M_s)| = |M_s|/p = p^{n+sd-1}$ for each $s \in \mathbb{N}_0$. In conclusion, $|K_s| = |H^1(R, M_0)|p^{n+sd-1}$ for every s, and therefore $|K_{s+1}|/|K_s| = p^d$, as claimed.

Second, we prove $|J_{s+1}| = p^e|J_s|$ for $s \in \mathbb{N}_0$. By the Orbit-Stabiliser Theorem, $|J_s| = |\mathrm{Comp}(R, M_s)|/|\mathrm{Comp}(R, M_s) \cdot \xi^s(\varphi)|$ where $\mathrm{Comp}(R, M_s) \cdot \xi^s(\varphi)$ denotes the $\mathrm{Comp}(R, M_s)$-orbit of $\xi^s(\varphi)$. The proof of Theorem 7.3 shows that $|\mathrm{Comp}(R, M_s) \cdot \xi^s(\varphi)| = |\mathrm{Comp}(R, M_0) \cdot \varphi|$ is independent of s: indeed, the respective cohomology groups $H^2(R, M_s)$ and $H^2(R, M_0)$ are isomorphic, and Step (b5) in the proof shows that the orbits in these groups under the compatible pairs correspond to each other. Therefore the claim $|J_{s+1}| = p^e|J_s|$ follows if we establish that

$$|\mathrm{Comp}(R, M_{s+1})| = p^e|\mathrm{Comp}(R, M_s)|$$

for all $s \in \mathbb{N}_0$. The structure of $\mathrm{Comp}(R, M_s)$ is discussed in Theorem 5.26 and the material preceding it. For a fixed $\alpha \in \mathrm{Aut}(R)$ let $H_{\alpha,s}$ be the image of the map $\sigma_{\alpha,s}\colon \mathrm{Hom}_R(T, T_\alpha) \to \mathrm{Hom}_R(T, (M_s)_\alpha)$ that is induced by the projection $T \to M_s$; here T_α denotes the R-module T where $g \in R$ acts as $\alpha(g)$. The kernel of $\sigma_{\alpha,s}$ is $L_{\alpha,s} = \mathrm{Hom}_R(T, (T_{n+sd})_\alpha)$, and so $|H_{\alpha,s+1}|/|H_{\alpha,s}| = [L_{\alpha,s} : L_{\alpha,s+1}]$. Since $T_{n+id} = p^i T_n$ for all i, it follows that $[L_{\alpha,s} : L_{\alpha,s+1}] = [L_{\alpha,0} : L_{\alpha,1}]$ is independent of $s \in \mathbb{N}_0$. Since this holds for all $\alpha \in \mathrm{Aut}(R)$, it follows from Theorem 5.26 that $|\mathrm{Comp}(R, M_{s+1})|/|\mathrm{Comp}(R, M_s)|$ is independent of s, so it equals p^e as defined prior to the proposition. This proves $|J_{s+1}| = p^e|J_s|$, and completes the proof. □

7.5 Applications: Schur multipliers, character theory, and more

We survey some other results on coclass families and mention some associated open problems.

7.5.1 Schur multipliers of p-groups

The *Schur multiplier* $M(G)$ of a group G is its second homology group $H_2(G,\mathbb{Z})$. Introduced by Schur in 1911 as part of his study of projective representations, it is now a central tool in group theory with applications ranging from central extensions to Galois theory. Hopf's Formula [72, (11.4.15)] shows that $M(G) \cong (F' \cap R)/[F,R]$ where $G = F/R$ is given as a quotient of a free group F. Note that $M(G)$ is abelian; if G is a finite p-group, then $M(G)$ is a finite abelian p-group; we refer to [50] for details.

Coclass theory has provided interesting insights into the structure of the Schur multipliers of finite p-groups. For example, Eick & Feichtenschlager [28] proved the following for the Schur multipliers of the groups in a coclass family.

Proposition 7.9 *Let $(G_s \mid s \in \mathbb{N}_0)$ be a coclass family. There exist $r_1, \ldots, r_m, t \in \mathbb{N}_0$ and $s_1, \ldots, s_m \in \mathbb{Z}$ such that for all $s \geqslant t$*

$$M(G_s) \cong C_{p^{r_1 s+s_1}} \times \ldots \times C_{p^{r_m s+s_m}}.$$

The proof in [28] applied Hopf's Formula to the parametrised presentation (Theorem 7.6) describing the groups in the family $(G_s \mid s \in \mathbb{N}_0)$; this is a rather technical symbolic computation and we omit the details here. Instead, we emphasise the following corollary which is analogous to Proposition 7.8.

Corollary 7.10 *If $(G_s \mid s \in \mathbb{N}_0)$ is a coclass family, then there exist $t, y \in \mathbb{N}_0$ such that for all $s \geqslant t$ the rank of $M(G_s)$ is independent of s and $|M(G_{s+1})| = p^y |M(G_s)|$.*

Recall that S is the infinite pro-p-group associated with the maximal coclass tree $\mathcal{T}$ in $\mathcal{G}(p,r)$. The structure of the finitely generated abelian pro-p-group $M(S)$ and its relation to $M(G)$ for G in $\mathcal{T}$ were investigated by Eick [26, 27]; some of the results can be summarised as follows.

Proposition 7.11 *Using this notation we obtain the following.*

a) *If $|M(S)| = \infty$ (which holds for $p > 2$), then for every $t \in \mathbb{N}$ there are finitely many groups G in $\mathcal{T}$ with $|M(G)| \leqslant t$.*
b) *If $|M(S)| < \infty$, then there exists $t \in \mathbb{N}$ such that $|M(G)| \leqslant t$ for all $G \in \mathcal{T}$.*

Recent work on Schur multipliers of groups of a given coclass include that of Moravec [64] and Joshi & Sarkar [48].

7.5.2 Character theory for finite p-groups

Couson [10, 11] has investigated the irreducible characters of the groups in a coclass family. We state his result as follows.

Proposition 7.12 *Let* $(G_s \mid s \in \mathbb{N}_0)$ *be a coclass family. There exists an integer* b *such that every irreducible character of every group* G_s *has degree at most* p^b*. Further, for each* $l \in \{0, \dots, b\}$ *there exists a polynomial* $f_l(\mathbf{x}) \in \mathbb{Q}[\mathbf{x}]$ *such that the number of irreducible characters of degree* p^l *of* G_s *is* $f_l(p^s)$*, provided that* s *is sufficiently large.*

This theorem supports the conjecture that the irreducible characters of the groups in a coclass family share structural similarities. We formulate this as follows, even though the statement of the conjecture is vague as it does not clarify the process of *how* the characters of G_{s+1} can be determined from those of G_s.

Conjecture 7.13 Let $(G_s \mid s \in \mathbb{N}_0)$ be a coclass family. There exists $t \in \mathbb{N}$ such that for all $s \geqslant t$ the ordinary character table of G_{s+1} can be determined from the character table of G_s.

7.5.3 Cohomology rings of finite p-groups

Let G be a finite p-group and let M be a commutative ring, considered as a trivial G-module. The associated cohomology ring is the direct sum

$$H^*(G, M) = \bigoplus_{i \in \mathbb{N}_0} H^i(G, M)$$

together with a *cup-product* defined by maps $H^i(G, M) \times H^j(G, M) \to H^{i+j}(G, M)$ for $i, j \in \mathbb{N}_0$; we refer to Carlson [8] and Carlson et al. [7] for extensive details.

Cohomology rings provide a wealth of open problems in the theory of finite p-groups. A recent result of Symonds [80] established that the infinitely many groups in $\mathcal{G}(p, r)$ yield only finitely many mod-p cohomology rings (with module $M = \mathbb{Z}/p\mathbb{Z}$), up to isomorphism. Eick & Green [29, 30] proved that the *Quillen categories* of almost all groups in a coclass family coincide. These results support the following conjecture.

Conjecture 7.14 Let $(G_s \mid s \in \mathbb{N}_0)$ be a coclass family. There exists $t \in \mathbb{N}$ such that $H^*(G_s, \mathbb{Z}/p\mathbb{Z}) \cong H^*(G_{s+1}, \mathbb{Z}/p\mathbb{Z})$ for all $s \geqslant t$.

Not much is known about the cohomology rings for other modules. For integral modules, we propose the following conjecture.

Conjecture 7.15 Let $(G_s \mid s \in \mathbb{N}_0)$ be a coclass family. There exists $t \in \mathbb{N}$ such that for all $s \geqslant t$ the integral cohomology rings $H^*(G_s, \mathbb{Z})$ can be described by a single parametrised presentation with parameter s.

Chapter 8
The skeletons of a coclass tree

The groups on the mainline of a maximal coclass tree $\mathcal{T}$ in a graph $\mathcal{G}(p,r)$ can be determined as quotients $S/\gamma_i(S)$ of the infinite pro-p-group S associated with $\mathcal{T}$. In this chapter we describe a method to construct groups in $\mathcal{T}$ that are not on the mainline. For large enough j, these groups form a subtree (the *skeleton*) $\mathcal{S}_j$ of the branch $\mathcal{B}_j$ in $\mathcal{T}$; this subtree has the same root as $\mathcal{B}_j$.

The construction described here is a new approach to the *constructible groups* defined by Leedham-Green & McKay [57, Definition 8.4.9], and we show that the groups in the skeletons coincide with the constructible groups. Importantly, [57, Theorems 1.3.7 and 11.3.9] state that there is a constant $c = c(p,r)$ such that every group in $\mathcal{B}_j$ has distance at most c from a group in $\mathcal{S}_j$. This justifies the name 'skeleton', because $\mathcal{S}_j$ exhibits the broad shape of the branch $\mathcal{B}_j$. We also show that the sequence of branches $\mathcal{B}_j$ in $\mathcal{T}$ has unbounded depth if and only if the sequence of skeletons $\mathcal{S}_j$ has unbounded depth. The similar result can be observed for the widths.

Recall that S is a uniserial p-adic pre-space group, and therefore an extension of $T = (\mathbb{Z}_p^d, +)$ by a finite p-group P. Our construction makes a case distinction depending on whether S is a split extension of P by T, or not. We show that for $p = 2$ the skeleton $\mathcal{S}_j$ only contains the root of $\mathcal{B}_j$, which is why the discussion in this chapter mostly assumes that p is an odd prime. As a preliminary step, we discuss the exterior square $T \wedge T$.

8.1 The exterior square and homomorphisms

Let T be an R-module, where R is a commutative unital ring. The *tensor square* $T \otimes T$ is the quotient of the free R-module generated by abstract symbols $s \otimes t$ (for $s, t \in T$) modulo the submodule generated by the relations

$$(u + \alpha v) \otimes t = (u \otimes t) + \alpha(v \otimes t)$$
$$u \otimes (s + \alpha t) = (u \otimes s) + \alpha(u \otimes t)$$

H. Dietrich and B. Eick, *Coclass graphs of p-groups*, Springer Monographs in Mathematics, https://doi.org/10.1007/978-3-032-17506-9_8

for all $u, v \in T$ and $\alpha \in R$. The R-module $T \otimes T$ is uniquely determined (up to isomorphism) by the following universal property: for every R-bilinear map φ from $T \times T$ into an R-module S, there is a unique R-linear map $\varphi' \colon T \otimes T \to S$ such that $(u, v)^\varphi = (u \otimes v)^{\varphi'}$ for all $u, v \in T$.

The *exterior square* $T \wedge T$ is the quotient of $T \otimes T$ modulo the submodule I generated by all $u \otimes u$ with $u \in T$. We write $u \wedge v = (u \otimes v) + I$; note that $u \wedge v = -(v \wedge u)$.

Lemma 8.1 *Let T be a free $\mathbb{Z}_p$-module with basis $\{t_1, \ldots, t_d\}$. A $\mathbb{Z}_p$-basis of $T \wedge T$ is $\{t_i \wedge t_j \mid 1 \leqslant i < j \leqslant d\}$.*

Proof Elements u and v of T can be written uniquely as $u = \alpha_1 t_1 + \ldots + \alpha_d t_d$ and $v = \beta_1 t_1 + \ldots + \beta_d t_d$ with $\alpha_i, \beta_j \in \mathbb{Z}_p$. A straightforward calculation shows that

$$u \wedge v = \sum_{i<j} (\alpha_i \beta_j - \alpha_j \beta_i)(t_i \wedge t_j).$$

Thus, the $t_i \wedge t_j$ with $i < j$ generate $T \wedge T$ as $\mathbb{Z}_p$-module. The linear independence follows similarly. □

The exterior square occurs naturally in the context of group commutators. Let G be a group of class 2, so $G > \gamma_2(G) > \gamma_3(G) = \{1\}$. The group $Q = G/\gamma_2(G)$ is abelian, hence a $\mathbb{Z}$-module, and so we can construct $Q \wedge Q$. Now consider the map $\delta \colon Q \wedge Q \to \gamma_2(G)$ defined by $g\gamma_2(G) \wedge h\gamma_2(G) \mapsto [g, h]$. Since $\gamma_2(G)$ is central, it follows that δ is a well-defined epimorphism.

8.1.1 Homomorphisms

We consider a finite p-group P and uniserial P-module $T \cong (\mathbb{Z}_p^d, +)$ with series $T = T_0 > T_1 > \ldots$. The P-action on T allows us to furnish $T \wedge T$ with a P-module structure with diagonal action. This action extends naturally to a $\mathbb{Z}_p P$-module structure. For $j \geqslant 0$ we denote by $T_j \wedge T$ the $\mathbb{Z}_p$-submodule of $T \wedge T$ generated by all $s \wedge t$ with $s \in T_j$ and $t \in T$. We recall from Remark 1.21 that every P-module homomorphism $T \wedge T \to T$ is automatically a $\mathbb{Z}_p$-module homomorphism, and therefore a $\mathbb{Z}_p P$-module homomorphism.

We start with the following observation; we denote by d the $\mathbb{Z}_p$-dimension of T. Recall that $d = d_s = (p-1)p^{s-1}$ for some s by Proposition 4.14.

Lemma 8.2 *Let $f \colon T \wedge T \to T$ be a non-trivial $\mathbb{Z}_p P$-module homomorphism. Then $f(T \wedge T) = T_j$ for some $j \in \mathbb{N}_0$. If $j \leqslant m \leqslant 2j - d$, then f induces a well-defined $\mathbb{Z}_p P$-module homomorphism $(T/T_m) \wedge (T/T_m) \to T_j/T_m$.*

Proof Since f is a $\mathbb{Z}_p P$-module homomorphism, the image of a P-submodule of $T \wedge T$ under f is an ideal of the ring $\mathbb{Z}_p[\theta]$. Proposition 4.4 shows that $f(T \wedge T) = T_j$ and $f(T_j \wedge T) = T_k$ for some $k \geqslant j$. We write $j = ad + b$ with $a \geqslant 0$ and $0 \leqslant b < d$; note that $(a+1)d \geqslant j$. Then $T_j = p^a T_b$ and

$$T_k = f(T_j \wedge T) = p^a f(T_b \wedge T) \leqslant p^a T_j = T_{j+ad} \leqslant T_{2j-d},$$

which shows that $k \geqslant 2j - d$. Thus, if $j \leqslant m \leqslant 2j - d$, then $T_k \leqslant T_m \leqslant T_j$, and therefore $f(T_m \wedge T) \leqslant f(T_j \wedge T) \leqslant T_{2j-d} \leqslant T_m$. This implies the claim. □

We write $\mathrm{Hom}_P(T \wedge T, T)$ for the group of P-homomorphisms from $T \wedge T$ to T; this is a $\mathbb{Z}_p$-module with the usual addition and scalar multiplication. We recall the following from [57, Theorem 11.4.4].

Proposition 8.3 *Let $p > 2$. The group $\mathrm{Hom}_P(T \wedge T, T)$ is a free $\mathbb{Z}_p$-module of rank at least $(p-3)/2$. This bound is achieved if P acts as W_s on T, where W_s is a Sylow p-Sylow subgroup of $\mathrm{GL}_{d_s}(\mathbb{Z}_p)$ (see Definition 4.13), and T has $\mathbb{Z}_p$-dimension d_s.*

For $p = 2$ we observe the following.

Lemma 8.4 *If $p = 2$ and P acts faithfully on T, then $\mathrm{Hom}_P(T \wedge T, T)$ is trivial.*

Proof Proposition 4.8 shows that P acts irreducibly on T. Since P is a finite 2-group, there exists an central involution $g \in Z(P)$, see Proposition 1.4. Acting with g therefore induces a $\mathbb{Z}_2P$-module isomorphism $T \to T$. Schur's Lemma [47, Lemma 1.5] now shows that g acts non-trivially by multiplication by a scalar of order 2, hence, by multiplication by -1. Thus, if $f\colon T \wedge T \to T$ is a $\mathbb{Z}_2P$-homomorphism, then for every $a, b \in T$

$$-f(a \wedge b) = f(a \wedge b)^g = f(a^g \wedge b^b) = f(-a \wedge -b) = f(a \wedge b).$$

It follows that f is the 0-homomorphism. □

8.1.2 Homomorphisms for cyclic *P*

Leedham-Green & McKay [53–57] extensively studied the case when P is cyclic. Motivated by Lemma 8.4, we assume $p > 2$.

Let θ be a primitive p^s-th root of unity over $\mathbb{Q}_p$. Let $P = \langle \theta \rangle$ and $T = (\mathbb{Z}_p[\theta], +)$. Then P acts by multiplication on T, and $T \cong (\mathbb{Z}_p^d, +)$ where $d = (p-1)p^{s-1}$. Recall from Section 3.4.1 that T has a unique maximal ideal $\mathfrak{p}$ generated by $\theta - 1$ and that P acts uniserially on T with series $T = T_0 > T_1 > \ldots$ where each $T_n = (\mathfrak{p}^n, +)$. A $\mathbb{Z}_p$-basis of T is given by $B = \{1, \theta, \ldots, \theta^{d-1}\}$.

Working with the group ring $\mathbb{Z}_pP$ and the ring $\mathbb{Z}_p[\theta]$ leads to ambiguous notation. For example, in $\mathbb{Z}_p[\theta]$

$$1 + \theta^{p^{s-1}} + \theta^{2p^{s-2}} + \ldots + \theta^{(p-1)p^{s-1}} = 0, \tag{8.1}$$

but in the group ring $\mathbb{Z}_pP$ the element $1 + \theta^{p^{s-1}} + \theta^{2p^{s-1}} + \ldots + \theta^{(p-1)p^{s-1}}$ is not 0. In general, $T \wedge T$ is a $\mathbb{Z}_p$-module, but not a $\mathbb{Z}_p[\theta]$-module with diagonal action.

Remark 8.5 If $p^s = 3$, then $\theta^2 = 1 + \theta$ and $T \wedge T$ has $\mathbb{Z}_3$-basis $\{1 \wedge \theta\}$. If $f\colon T \wedge T \to T$ is a $\mathbb{Z}_3P$-module homomorphism, then

$$f(1 \wedge \theta)^\theta = f(\theta \wedge \theta^2) = f((1+\theta) \wedge \theta) = f(1 \wedge \theta).$$

Since $0 \in T$ is the only fixed point under P, it follows that $f(1 \wedge \theta) = 0$. Thus, there is no non-trivial $\mathbb{Z}_3 P$-module homomorphism $T \wedge T \to T$. We see below that the situation is different when $p^s \neq 3$.

Recall from (3.2) that the only relations satisfied by θ are those induced by (8.1). In the following, write $\mathbb{K} = \mathbb{Q}_p(\theta)$. For an integer i coprime to p let σ_i be the automorphism of $\mathbb{K}$ that fixes $\mathbb{Q}_p$ element-wise and maps θ to θ^i. For an integer $a \not\equiv 0, 1 \bmod p$, let the $\mathbb{Q}_p$-linear homomorphisms $\mathbb{S}_a \colon \mathbb{K} \wedge \mathbb{K} \to \mathbb{K}$ be defined by

$$\mathbb{S}_a \colon \mathbb{K} \wedge \mathbb{K} \to \mathbb{K}, \quad \mathbb{S}_a(u \wedge v) = \sigma_a(u)\sigma_{1-a}(v) - \sigma_a(v)\sigma_{1-a}(u). \tag{8.2}$$

A short calculation shows that each $\mathbb{S}_a \in \mathrm{Hom}_{\mathbb{Q}_p P}(\mathbb{K} \wedge \mathbb{K}, \mathbb{K})$. Leedham-Green & McKay [57, Theorem 11.4.1] observed that these homomorphisms define a basis of $\mathrm{Hom}_{\mathbb{Q}_p P}(\mathbb{K} \wedge \mathbb{K}, \mathbb{K})$; we include the proof here.

Proposition 8.6 *The set* $\{\mathbb{S}_a \mid a = 2, \ldots, (p^s - 1)/2,\ a \not\equiv 0, 1 \bmod p\}$ *forms a* $\mathbb{K}$*-basis of* $\mathrm{Hom}_{\mathbb{Q}_p P}(\mathbb{K} \wedge \mathbb{K}, \mathbb{K})$ *of size* $(p^s - 1)/2 - p^{s-1}$.

Proof We first show that $\Sigma = \{\sigma_a \mid a = 1, \ldots, p^s - 1,\ p \nmid a\}$ is a $\mathbb{K}$-linearly independent set in the ring of $\mathbb{Q}_p$-endomorphisms $\mathbb{K} \to \mathbb{K}$. If not, then there is a shortest linear combination $k_1 s_1 + \ldots + k_t s_t = 0$ where $s_1, \ldots, s_t \in \Sigma$ are distinct and $k_1, \ldots, k_t \in \mathbb{K}$ are all nonzero. Suppose, for a contradiction, that $t > 1$. Each s_i has the form $s_i = \sigma_{j_i}$ for some j_i, that is, $s_i(\theta) = \theta^{j_i}$. If μ_i denotes multiplication by θ^i, then each $s_\ell \circ \mu_1 = \mu_{j_\ell} \circ s_\ell = \theta^{j_\ell} s_\ell$. Thus, composing $k_1 s_1 + \ldots + k_t s_t = 0$ from the left with μ_{i_1} and from the right with μ_1, respectively, yields

$$k_1 \theta^{i_1} s_1 + \ldots + k_t \theta^{i_1} s_t = 0 = k_1 \theta^{i_1} s_1 + \ldots + k_t \theta^{i_t} s_t.$$

Subtracting both equations yields $k_2(\theta^{i_1} - \theta^{i_2})s_2 + \ldots + k_t(\theta^{i_1} - \theta^{i_t})s_t = 0$. Since the differences $\theta^{i_1} - \theta^{i_j}$ are nonzero for each $j = 2, \ldots, t$, this is a shorter non-trivial linear combination, contradicting our initial choice. Thus, $t = 1$. Now $k_1 s_1 = 0$ with $k_1 \neq 0$ contradicts that s_1 is an automorphism. Hence, Σ is a linearly independent set; this argument is sometimes called Dedekind's Independence Theorem.

Recall from Proposition 3.14 that $\mathbb{K}$ has $\mathbb{Q}_p$-dimension $d = (p-1)p^{s-1}$, so the $\mathbb{K}$-space of $\mathbb{Q}_p$-endomorphisms $\mathbb{K} \to \mathbb{K}$ has dimension d. Since $|\Sigma| = d$, the set Σ is a basis of this space. Write $\mathcal{I} = \{i \in \mathbb{N} \mid 1 \leqslant i \leqslant p^s - 1 \text{ and } p \nmid i\}$. It follows that the set of all $\sigma_a \otimes \sigma_b$ with $a, b \in \mathcal{I}$ forms a $\mathbb{K}$-basis of the space of $\mathbb{Q}_p$-homomorphisms $\mathbb{K} \otimes \mathbb{K} \to \mathbb{K}$. To determine a basis of $\mathrm{Hom}_P(\mathbb{K} \otimes \mathbb{K}, \mathbb{K})$, we investigate when an arbitrary $f = \sum_{a,b \in \mathcal{I}} k_{a,b} \sigma_a \otimes \sigma_b$ commutes with the action of P. It suffices to check when f commutes with multiplication by θ, which holds if and only if

$$\sum_{a,b \in \mathcal{I}} k_{a,b} \theta^{a+b} \sigma_a \otimes \sigma_b = \sum_{a,b \in \mathcal{I}} k_{a,b} \theta \sigma_a \otimes \sigma_b,$$

if and only if $k_{a,b}(\theta^{a+b} - \theta) = 0$ for all a, b, if and only if p^s divides $a + b - 1$, if and only if $b \equiv 1 - a \bmod p^s$. Thus, a basis of the $\mathbb{K}$-space $\mathrm{Hom}_{\mathbb{Q}_p P}(\mathbb{K} \otimes \mathbb{K}, \mathbb{K})$ is given by the set of homomorphisms $\sigma_a \otimes \sigma_{1-a}$ for $a \in \{1, \ldots, p^s - 1\}$ with

$a \not\equiv 0, 1 \bmod p^s$. The last step is to investigate when $f = \sum_a k_a \sigma_a \otimes \sigma_{1-a}$ induces an element in $\mathrm{Hom}_{\mathbb{Q}_p P}(\mathbb{K} \wedge \mathbb{K}, \mathbb{K})$. Since f is a $\mathbb{Q}_p P$-homomorphism, if it induces a $\mathbb{Q}_p P$-homomorphism $\mathbb{K} \wedge \mathbb{K} \to \mathbb{K}$, then f must map $(1 \otimes \theta) - (\theta \otimes 1)$ to 0. The latter holds if and only if $\sum_a k_a \theta^{1-a} = \sum_a k_a \theta^a$, if and only if $\sigma_a(k_a - k_{1-a})\theta^a = 0$, if and only if $k_{1-a} = -k_a$, if and only if f is a linear combination of the $\mathbb{S}_a$. □

As a corollary, we obtain the following for the restriction to $T = \mathbb{Z}_p[\theta]$; this is [57, Corollary 11.4.2].

Proposition 8.7 $\mathrm{Hom}_P(T \wedge T, T)$ *is a free* $\mathbb{Z}_p[\theta]$*-module of rank* $(p^s - 2p^{s-1} - 1)/2$.

We now further investigate the structure of $\mathrm{Hom}_P(T \wedge T, T)$. Every $\mathbb{Z}_p P$-homomorphism $T \wedge T \to T$ extends linearly to a $\mathbb{Q}_p P$-homomorphism $\mathbb{K} \wedge \mathbb{K} \to \mathbb{K}$. Conversely, every $\mathbb{S}_a$ restricts to a $\mathbb{Z}_p P$-homomorphism $T \wedge T \to T$, which, by abuse of notation, we will also denote by $\mathbb{S}_a$. However, the latter is not surjective onto T. If $i < j$, then $\mathbb{S}_a(\theta^i \wedge \theta^j) = \theta^i \mathbb{S}_a(1 \wedge \theta^{j-i})$, and for every $k \in \mathbb{N}$

$$\mathbb{S}_a(1 \wedge \theta^k) = \theta^{ka}(\theta^{k(1-2a)} - 1) \in T_1.$$

Note that $\theta^{k(1-2a)} - 1 = (\theta - 1)(1 + \theta + \ldots + \theta^{k(1-2a)-1}) \equiv (\theta - 1)k(1 - 2a) \bmod T_2$. Thus, if $2a \not\equiv 1 \bmod p$, then $\mathbb{S}_a(T \wedge T)$ is contained in T_1, but not in T_2, and therefore $\mathbb{S}_a(T \wedge T) = T_1$ by Proposition 4.4. This shows that for $2a \not\equiv 1 \bmod p$, the map $(\theta - 1)^{-1}\mathbb{S}_a$ is a surjective $\mathbb{Z}_p P$-homomorphism $T \wedge T \to T$. However, if $a = (cp+1)/2$ for some (necessarily odd) $c \in \{1, \ldots, p^{s-1} - 1\}$, then $\mathbb{S}_a(T \wedge T) < T_1$.

Lemma 8.8 *A* $\mathbb{Z}_p$*-basis of* $T \wedge T$ *is given by* $\{\theta^i \wedge \theta^j \mid 0 \leqslant i < j < d\}$*, and* $T \wedge T$ *is generated as* $\mathbb{Z}_p P$*-module by* $\{1 \wedge \theta^i \mid i = 1, \ldots, (p^s - 1)/2\}$.

Proof The first claim follows since $\{1, \theta, \ldots, \theta^{d-1}\}$ is a $\mathbb{Z}_p$-basis of T, see (8.1) and Lemma 8.1. Since the generator θ of P acts diagonally on $x \wedge y \in T \wedge T$ by multiplication by θ, that is, $(x \wedge y)^\theta = (\theta x) \wedge (\theta y)$, the claim follows from the observation that $(\theta^i \wedge \theta^j) = (1 \wedge \theta^{j-i})^{(\theta^i)}$ and $1 \wedge \theta^j = -(1 \wedge \theta^{p^s - j})^{(\theta^j)}$ for $i < j$. □

For $j \in \mathbb{N}$ denote by V_j the $\mathbb{Z}_p P$-submodule of $T \wedge T$ generated by $1 \wedge \theta^j$. The previous lemma shows that $T \wedge T = V_1 + \ldots + V_{(p^s - 1)/2}$, but this sum is not direct. We first look at the P-orbit of $1 \wedge \theta^i$. Note that if $p = 3$ and $s = 1$, then $(1 \wedge \theta)^\theta = \theta \wedge \theta^2 = \theta \wedge (-1 - \theta) = 1 \wedge \theta$, that is, the P-orbit of $1 \wedge \theta$ has size 1; the next lemma generalises this observation.

Lemma 8.9 *Let* $i, j \in \{1, \ldots, (p^s - 1)/2\}$.

a) *The* P*-orbits of* $1 \wedge \theta^i$ *and* $1 \wedge \theta^j$ *coincide if and only if* $i = j$.
b) *The* P*-orbit of* $1 \wedge \theta^i$ *has size* p^s*, unless* $p = 3$ *and* $s > 1$ *and* $i = 3^{s-1}$ *in which case the* P*-orbit has size* 3^{s-1}.

Proof a) The orbits coincide if and only if $1 \wedge \theta^j = (1 \wedge \theta^i)^{\theta^k} = \theta^k \wedge \theta^{i+k}$ for some $\theta^k \in P$ with $k \in \{0, 1, \ldots, p^s - 1\}$. If $k \geqslant d$, then we can swap i and j and act with θ^{-k} instead of θ^k; thus, we can assume that $k \in \{1, \ldots, d - 1\}$. If $(i + k) \bmod p^s < d$, then Lemma 8.8 forces $k = 0$ and $(i + k) \bmod p^s = j$, so $i = j$ as claimed. It remains to consider the case $(i+k) \bmod p^s \geqslant d$, say $(i+k) \bmod p^s = d+\ell$ with $\ell \in \{0, \ldots, p^{s-1}-1\}$. With (8.1) we rewrite $1 \wedge \theta^j = \theta^k \wedge \theta^{i+k}$ as

$$(*) \quad 1 \wedge \theta^j = \theta^\ell(1 + \theta^{p^{s-1}} + \ldots + \theta^{(p-2)p^{s-1}}) \wedge \theta^k.$$

All exponents of θ in this expression lie in $\{0, \ldots, d-1\}$, and Lemma 8.8 shows that $(*)$ holds only if $p = 3$ and $\theta^\ell \wedge \theta^k + \theta^{3^{s-1}} \wedge \theta^k = 1 \wedge \theta^j$, that is, $k = j = 3^{s-1}$ and $\ell = 0$. Now $(i+k) \bmod p^s = d$ becomes $(i + 3^{s-1}) \bmod 3^s = 2 \cdot 3^{s-1}$, and therefore $i = 3^{s-1} = j$, as claimed.

b) By the Orbit-Stabiliser Theorem, the size of the orbit is $|P| = p^s$ divided by the size of the stabiliser $U = \mathrm{Stab}_P(1 \wedge \theta^i)$. If there is a non-trivial $\theta^k \in P$ that maps $1 \wedge \theta^i$ to itself, then the proof of a) shows that this is possible if and only if $p = 3$ and $i = 3^{s-1}$ and $k = 3^{s-1}$. In this case, $|U| = |\theta^{3^{s-1}}| = 3$, and so the P-orbit has size 3^{s-1}. □

While the P-orbits of different $1 \wedge \theta^i$ are (obviously) disjoint, this is not true for the $\mathbb{Z}_p$-submodules generated by these orbits. In fact, determining the precise structure of $T \wedge T$ as $\mathbb{Z}_pP$-module is a delicate matter, which is why we focus on the special case $P \cong C_p$. This case was also discussed in [57, Section 8], and it will play a crucial role in Chapter 9.

Proposition 8.10 *Let $P = \langle\theta\rangle \cong C_p$, $T = (\mathbb{Z}_p[\theta], +)$, and $\kappa = \theta - 1$.*

- a) *$T \wedge T = C \oplus Z$ where C is a free $\mathbb{Z}_pP$-module of rank $(p-3)/2$ and Z is the P-invariant free $\mathbb{Z}_p$-module generated by $z = \sum_{0 \leqslant i < j \leqslant p-2} \theta^i \wedge \theta^j$.*
- b) *The $\mathbb{Z}_pP$-module C is generated freely by $\{1 \wedge \theta^i \mid i = 1, \ldots, (p-3)/2\}$.*
- c) *The $\mathbb{Z}_pP$-module C is generated freely by $\{\kappa^i \wedge \kappa^{i-1} \mid i = 1, \ldots, (p-3)/2\}$.*

Proof Recall that $d = p - 1$, so $\theta^d = -1 - \theta - \ldots - \theta^{-d-1}$.

a,b) Lemma 8.8 shows that $T \wedge T$ has $\mathbb{Z}_p$-basis $\mathcal{B} = \{\theta^i \wedge \theta^j \mid 0 \leqslant i < j < d\}$, and that $T \wedge T$ is generated as $\mathbb{Z}_pP$-module by $\{1 \wedge \theta, \ \ldots, \ 1 \wedge \theta^{d/2}\}$. If $p = 3$, then $(1 \wedge \theta)^\theta = \theta \wedge \theta^2 = 1 \wedge \theta$, and the claim follows. We now consider $p > 3$.

Let $i \in \{1, \ldots, d/2\}$. A short calculation shows that the $\mathbb{Z}_pP$-submodule generated by $1 \wedge \theta^i$ has $\mathbb{Z}_p$-dimension p, with basis

$$\mathcal{C}_i = \{1 \wedge \theta^i, \ \ldots, \ \theta^{d-i} \wedge \theta^d, \ 1 \wedge \theta^{d-i+1}, \ \ldots, \ \theta^{i-1} \wedge \theta^d\}.$$

Let C be the $\mathbb{Z}_p$-space spanned by $\mathcal{C} = \mathcal{C}_1 \cup \ldots \cup \mathcal{C}_{d/2-1}$; note that $\mathcal{C}$ is a disjoint union, see also Lemma 8.9. We show that $\mathcal{C}$ is a $\mathbb{Z}_p$-basis of C, so C is a free $\mathbb{Z}_pP$-module of rank $d/2 - 1 = (p-3)/2$. Note that

$$\mathcal{C} = (\mathcal{B} \cap \mathcal{C}) \cup \{1 \wedge \theta^d, \ldots, \theta^{d/2-2} \wedge \theta^d, \theta^{d/2+1} \wedge \theta^d, \ldots, \theta^{d-1} \wedge \theta^d\}$$

is a disjoint union and

$$\mathcal{B} \cap \mathcal{C} = \{\theta^i \wedge \theta^j \mid 0 \leqslant i < j < d \text{ and } j - i \notin \{d/2, d/2+1\}\}. \tag{8.3}$$

Using $1 + \theta + \ldots + \theta^d = 0$ to rewrite $\theta^j \wedge \theta^d \in \mathcal{C}$, a short calculation now shows that, modulo the $\mathbb{Z}_p$-span of $\mathcal{B} \cap \mathcal{C}$,

$$\theta^j \wedge \theta^d \equiv \begin{cases} (\theta^{j-d/2} \wedge \theta^j) + (\theta^{j-d/2-1} \wedge \theta^j) & (j \geqslant d/2+1) \\ (\theta^{j+d/2} \wedge \theta^j) + (\theta^{j+d/2+1} \wedge \theta^j) & (j \leqslant d/2-2). \end{cases} \tag{8.4}$$

The elements listed on the right side of Equation (8.4) do not occur in Equation (8.3), and since $\mathcal{B}$ is $\mathbb{Z}_p$-linearly independent, an inspection of Equation (8.4) reveals that $\mathcal{C}$ is a $\mathbb{Z}_p$-basis of C. This implies that C is generated by $\{1 \wedge \theta, \ldots, 1 \wedge \theta^{(p-3)/2}\}$ as a free $\mathbb{Z}_pP$-module of rank $(p-3)/2$.

We now consider the element z given in the statement of a). A short calculation using Equation (8.3) shows that, modulo the $\mathbb{Z}_p$-space spanned by $\mathcal{B} \cap \mathcal{C}$,

$$z \equiv \sum_{i=0}^{d/2-2} (\theta^i \wedge \theta^{d/2+i} + \theta^i \wedge \theta^{d/2+i+1}) + (\theta^{d/2-1} \wedge \theta^{d-1}).$$

Now (8.4) shows that $z \equiv (\theta^{d/2-1} \wedge \theta^{d-1}) \bmod C$, and therefore $z \notin C$. Since $T \wedge T$ has $\mathbb{Z}_p$-dimension $d(d-1)/2 = (p^2-3p)/2+1$ and C has $\mathbb{Z}_p$-dimension $p(d/2-1) = (p^2-3p)/2$, we deduce that $T \wedge T = C \oplus Z$. Lastly, note that

$$z^\theta = \sum_{1 \leqslant i < d} \theta^i \wedge \theta^d + \sum_{1 \leqslant i < j < d} \theta^i \wedge \theta^j = -(1 \wedge \theta^d) + \sum_{1 \leqslant i < j < d} \theta^i \wedge \theta^j = z,$$

so Z is fixed elementwise by P.

c) Let D be the $\mathbb{Z}_pP$-module generated by $\{\kappa^i \wedge \kappa^{i-1} \mid i = 1, \ldots, (p-3)/2\}$. We prove by induction on i that each $\mathcal{C}_i \subseteq D$. This shows that $C \leqslant D$. Since C is a free $\mathbb{Z}_pP$-module of rank $(p-3)/2$ and D is a $\mathbb{Z}_pP$-module with $(p-3)/2$ generators, $C = D$. Clearly, $1 \wedge \theta = 1 \wedge \kappa \in D$, so $\mathcal{C}_1 \subseteq D$. Now let $1 < i \leqslant (p-3)/2$ and assume that the $\mathbb{Z}_pP$-submodule E generated by $\mathcal{C}_1 \cup \ldots \cup \mathcal{C}_{i-1}$ lies in D. Expanding the powers κ^i and κ^{i-1}, we see that $\kappa^i \wedge \kappa^{i-1} = \theta^i \wedge (-1)^{i-1} + u$ for some $u \in E$. Since $\kappa^i \wedge \kappa^{i-1} \in D$ and $E \leqslant D$, we deduce that $1 \wedge \theta^i \in D$, hence $\mathcal{C}_i \subseteq D$, as claimed. □

8.2 Skeleton groups in the split case

In this section, p is an odd prime and S is an infinite pro-p-group of coclass r and dimension d such that $S = P \ltimes T$ for a translation subgroup T with point group P and uniserial series $T = T_0 > T_1 > \ldots$.

Definition 8.11 Let $f\colon T \wedge T \to T_j$ be a surjective $\mathbb{Z}_pP$-homomorphism and choose m such that $j \leqslant m \leqslant 2j-d$. Then T/T_m is a quotient of order p^m of the additive abelian group T and we define $T_{f,m} = (T/T_m, \circ_f)$ as the multiplicative group with underlying set T/T_m and multiplication

$$(u + T_m) \circ_f (v + T_m) = u + v + \tfrac{1}{2} f(u \wedge v) + T_m.$$

The *skeleton group* defined by f and m is

$$G_{f,m} = P \ltimes T_{f,m}. \tag{8.5}$$

Note that this construction is well-defined, c.f. Lemma 8.2: first, if we write $j = ad + b$ with $b \in \{0, \ldots, d-1\}$, then $T_j = p^a T_b$; since $j \leqslant m \leqslant 2j - d$

$$f(T_m \wedge T) \leqslant f(T_j \wedge T) = p^a f(T_b \wedge T) \leqslant p^a T_j = T_{j+ad} \leqslant T_{2j-d} \leqslant T_m.$$

Second, f is a $\mathbb{Z}_p P$-module homomorphism, so forming the semidirect product with P is possible and yields a group multiplication in $G_{f,m}$:

$$(g, u + T_m)(h, v + T_m) = (gh, u^h + v + \tfrac{1}{2} f(u^h \wedge v) + T_m).$$

The definition of $G_{f,m}$ depends on the choice of translation subgroup, which is implicitly encoded in the definition of f, but it is independent of the chosen complement P. If S is a p-adic space group, then its translation subgroup is unique, see Proposition 4.32. The following result is proved in a more general context in Theorem 8.17.

Proposition 8.12 *The group $G_{f,m}$ is a finite p-group of order $|P|p^m$ and coclass r. It is contained in the maximal coclass tree $\mathcal{T}$ associated with S and has depth $m - j$ in the branch with root of order $|P|p^j$.*

Hence, the mainline group S/T_j is isomorphic to the skeleton group $G_{f,j}$. Moreover, if $m \geqslant j + 1$, then $G_{f,m}/\gamma(G_{f,m}) \cong G_{f,m-1}$. Together with Proposition 8.12, this shows that the skeleton groups in $\mathcal{B}_j$ form a subtree with root S/T_j.

Corollary 8.13 *If there is a surjective $\mathbb{Z}_p P$-module homomorphism $f \colon T \wedge T \to T_j$ for some j, then the maximal coclass tree $\mathcal{T}$ associated with S has unbounded depth.*

Proof By construction, the path $G_{f,j}, G_{f,j+1}, \ldots, G_{f,2j-d}$ in Proposition 8.12 has length $j - d$ in the branch with root $G_{f,j}$; note that $(T/T_{j+1}, \circ_f)$ is not abelian, hence $G_{f,j+1}$ is not a mainline group, whereas $G_{f,j}$ is. Multiplication by p^k transforms f into a surjective $\mathbb{Z}_p P$-module homomorphism $p^k f \colon T \wedge T \to T_{j+kd}$, and this yields a path of length $j + kd - d$ in the branch with root $G_{p^k f, j+kd}$. The claim follows.□

If S has a cyclic point group P, then Lemma 8.4 and Proposition 8.7 show that the coclass tree associated with S has unbounded depth for $p \geqslant 3$ and $(p, r) \neq (3, 1)$.

8.3 Skeleton groups in the non-split case

Again, let S be an infinite pro-p-group of coclass r and dimension d. Let T be a translation subgroup of S with corresponding point group P and uniserial series $T = T_0 > T_1 > \ldots$. Recall that we assume $p > 2$.

Our approach to define skeleton groups in the case that S does not split over T follows ideas from [57, Chapter 8] as developed and expanded in [18]. The first step is to embed S as a finite-index subgroup of an infinite pro-p-group S^* with translation subgroup T^* such that $T^* \cong T$ and S^* splits over T^*. The second step is to construct skeleton groups for S^* as described in Section 8.2, and obtain skeleton groups for S as subgroups of those for S^*.

8.3.1 Split supergroups

We explain how S can be considered as a subgroup of a pro-p-group of finite coclass that splits over its point group, see also Corollary 5.30.

Theorem 8.14 *Let S be an infinite pro-p-group of coclass r, dimension d, translation subgroup T, and corresponding point group P. Extend the uniserial series $T = T_0 > T_1 > \ldots$ to the left by defining $T_{i-d} = p^{-1}T_i$ for all $i \in \mathbb{Z}$ via tensor extension of scalars. The group S embeds into an infinite pro-p-group S^* with translation subgroup $T^* = p^{-x}T$ for some $x \geqslant 0$ and point group P, such that $S^* = ST^*$ and $S \cap T^* = T$, and S^* is a split extension of P by T^*.*

Proof For convenience, we identify $T = (\mathbb{Z}_p^d, +)$ and $V = (\mathbb{Q}_p^d, +)$. We view S as an extension E_γ for some 2-cocycle $\gamma \in Z^2(P, T)$, that is, S has the underlying set $P \times T$ and multiplication

$$(g, s)(h, t) = (gh, s^h + t + \gamma(g, h)).$$

Letting $(g, t) \in S$ act on V as g, we consider the split extension $S \ltimes V$ and its normal subgroup N generated by $((1, -t), t)$ for all $t \in T$. The factor $H = (S \ltimes V)/N$ has a surjective projection $H \to P$, $((g, s), t)N \to g$, whose kernel $\{((1, 0), v)N \mid v \in V\}$ is isomorphic to V, that is, H is an extension of P by V. There are natural embeddings

$$S \to H, \quad (g, t) \mapsto ((g, t), 0)N, \quad \text{and}$$
$$V \to H, \quad v \mapsto ((1, 0), v)N,$$

that allow us to consider S and V as subgroups of H. As such, $H = SV$ and $S \cap V = T$. Since the additive group V is uniquely divisible by the p-power $|P|$, Proposition 5.2 shows that $H^2(P, V)$ is trivial, hence the extension H splits and so there is a complement $P^* \leqslant H$ to V in H, that is, $H = P^*V \cong P^* \ltimes V$. In particular, $P^* \cong P$, and for every $g \in P$ there exists $t_g \in V$ such that

$$P^* = \{((g, 0), t_g)N : g \in P\}. \tag{8.6}$$

In the factorisation $H = P^*V$, the image of $(g, 0) \in S$ under the embedding $S \to H$ is hv where $h = ((g, 0), t_g)N \in P^*$ and $v = ((1, 0), -t_g)N \in V$.

Let $x \in \mathbb{N}_0$ be minimal such that $t_g \in p^{-x}T$ for each $g \in P$. Replacing V by $T^* = T_{-xd}$ in the above construction yields the required group S^*. If we identify its underlying set with $\{(g, t) \mid g \in P, t \in T^*\}$, then the embedding $S \to S^*$ maps $(g, t) \mapsto (g, t)$, and a complement to T^* in S^* is given by $P^* = \{h_g \mid g \in P\}$ where each h_g is defined by $h_g = (g, t_g) \in S^*$. □

8.3.2 Skeleton groups and subgroups

As before, we write $S = P.T$ to indicate that S has a normal subgroup T with quotient group S/T isomorphic to P. As in Theorem 8.14, we extend the uniserial series of T to the left by setting $T_{i-d} = p^{-1}T_i$ for all $i \in \mathbb{Z}$, and consider the embedding of S in the pro-p-group $S^* = P.T^*$, where $T^* = p^{-x}T$ for a minimal non-negative x such that S^* is a split extension of P by T^*. To define skeleton groups for S, we embed S into S^*, construct skeleton groups for S^* as in Definition 8.11, and then translate these to groups that are defined for S. We now describe the three steps.

(1) We use the construction of Theorem 8.14 to embed S into the group S^* with $T^* \leqslant S^*$ such that $S^* = ST^*$ and $S \cap T^* = T$. Identifying the underlying sets of $S = P.T$ and $S^* = P.T^*$ with $\{(g,t) \mid g \in P, t \in T\}$ and with $\{(g,t) \mid g \in P, t \in T^*\}$, respectively, the embedding $\iota\colon S \to S^*$ maps each (g,t) to (g,t). By construction, T^* has a complement P^* in S^* isomorphic to P, that is, $S^* = P^* \ltimes T^*$ and P^* can be written as

$$P^* = \{h_g \mid g \in P\} \quad \text{with each } h_g = (g, t_g) \text{ for a suitable } t_g \in T^*.$$

Note that S^* is originally constructed as the non-split extension $S^* = P.T^*$ and now also as a split extension $S^* = P^* \ltimes T^*$; the element $(g,t) \in P.T^*$ corresponds to $(h_g, t - t_g) \in P^* \ltimes T^*$.

(2) In the following, let $j \geqslant \max\{2xd, d\}$ be an integer and write

$$j^* = j - 2xd.$$

Let $f\colon T \wedge T \to T_j$ be a surjective $\mathbb{Z}_pP$-homomorphism. It induces a surjective $\mathbb{Z}_pP$-homomorphism $T^* \wedge T^* \to T_{j^*}$, which, by abuse of notation, we also call f. A short calculation (similar to that after Definition 8.11) shows that if $m \in \mathbb{Z}$ satisfies $j^* \leqslant m \leqslant 2j^* + (x-1)d$, then $f(T_m \wedge T^*) \leqslant T_m$. Following Definition 8.11, we define the skeleton group

$$G^*_{f,m} = P^* \ltimes T^*_{f,m}$$

where $T^*_{f,m}$ is the group T^*/T_m whose multiplication is twisted by $f\colon T^* \wedge T^* \to T_{j^*}$, see (8.5).

(3) Let π be the natural projection from $G^*_{f,m}$ to S^*/T_{j^*} with kernel T_{j^*}/T_m. The inclusion $\iota\colon S \to S^*$ induces an embedding $S/T_{j^*} \to S^*/T_{j^*}$; to simplify the notation, we also denote it by ι. We now define

$$G_{P^*,f,m} \leqslant G^*_{f,m}$$

as the largest subgroup of $G^*_{f,m}$ such that $\pi(G_{P^*,f,m}) = \iota(S/T_{j^*})$. Recall that ι maps $(g, t + T_{j^*}) \in S/T_{j^*}$ to $(h_g, t - t_g + T_{j^*}) \in P^* \ltimes T^*/T_{j^*}$, and the full preimage of this element under π is $\{(h_g, t - t_g + s + T_m) \mid s \in T_{j^*}\}$. Since $T_{j^*} \leqslant T$ by our assumptions on j, the subgroup $G_{P^*,f,m}$ takes the form

$$G_{P^*,f,m} = \{(h_g, t - t_g + T_m) \mid g \in P, t \in T\} \leqslant P^* \ltimes T^*_{f,m}. \tag{8.7}$$

Since $P \cong \{h_g \mid g \in P\}$ and each h_g acts as g on T, we simplify the definition as follows.

Definition 8.15 Let S, P, T, x, and $P^* = \{h_g \mid g \in P\}$ with $h_g = (g, t_g)$ be as above. Let $f: T \wedge T \to T_j$ be a surjective $\mathbb{Z}_p P$-homomorphism such that $j^* = j - 2xd \geqslant 0$, and let m be such that $j^* \leqslant m \leqslant 2j^* + (x-1)d$. The *skeleton group* defined by f and m and the complement P^* is

$$G_{P^*,f,m} = \{(g, t - t_g + T_m) \mid g \in P, t \in T\} \leqslant P \ltimes T^*_{f,m}. \tag{8.8}$$

Remark 8.16 We continue with the previous notation.

a) The projection $G_{P^*,f,m} \to P$ is onto and has kernel $\{(1, t+T_m) \mid t \in T\} \cong T_{f,m}$, so the skeleton group $G_{P^*,f,m}$ is an extension of P by $T_{f,m}$.

b) If S splits, then $T = T^*$ and we choose $P = P^*$ and each $t_g = 0$. In this case, the skeleton group $G_{P^*,f,m}$ is that given in Definition 8.8, that is, $G_{f,m}$.

c) Recall that $P^* = \{h_g \mid g \in P\}$ with $h_g = (g, t_g)$, and if $g, k \in P$, then $h_{gk} = h_g h_k$. If $\gamma \in Z^2(P, T)$ is the 2-cocycle defining the extension S of P by T, then

$$h_{gk} = h_g h_k = (gk, t_g^k + t_k + \gamma(g,k)) \in P.T,$$

which implies that $t_{gk} = t_g^k + t_k + \gamma(g,k)$, and so $t_{gk} + T = t_g^k + t_k + T$. This can be used to verify directly that the set given in (8.7) forms a subgroup of $G^*_{f,m}$. It also shows that $\gamma \in Z^2(P,T)$ is a 2-coboundary in $B^2(P, T^*)$ defined by the map $P \to T^*$, $g \mapsto t_g$; c.f. Corollary 5.30.

We can now state the main result of this section; it generalises Proposition 8.12.

Theorem 8.17 *Let $S = P.T$ be an infinite pro-p-group of coclass r and dimension d, with translation subgroup T. As before, we embed S in $S^* = P.T^*$ with $T^* = p^{-x}T$ for some minimal non-negative x. Let c be the nilpotency class of P, let $k \in \mathbb{N}$ such that $\gamma_{c+1}(S) = T_k$, and let $j_0 = \max\{k, d+1\}$. Let $f: T \wedge T \to T_j$ be a surjective $\mathbb{Z}_p P$-homomorphism such that $j^* > j_0$, where $j^* = j - 2xd$. Let $G = G_{P^*,f,m}$ be a skeleton group where m is an integer with $j^* \leqslant m \leqslant 2j^* - (x-1)d$.*

a) *The group G is a finite p-group of coclass r that has S/T_{j^*} as a quotient.*

b) *The groups $G_{P^*,f,j^*}, G_{P^*,f,j^*+1}, \ldots, G_{P^*,f,2j^*+(x-1)d}$ form a path in $\mathcal{G}(p,r)$.*

c) *If $x = 0$, then $G = G_{f,m}$ lies in the branch with root S/T_j.*

Proof For $i \leqslant m$ write $N_i = \{(1, t + T_m) \mid t \in T_i\} \leqslant G$. Since $f(T^* \wedge T^*) = T_{j^*}$ and $f(T_{j^*} \wedge T^*) \leqslant T_m$, the group N_{j^*} is normal in S. It follows from Step **(3)** and Remark 8.16c) that

$$\alpha: S/T_{j^*} \to G/N_{j^*}, \quad (g, t + T_{j^*}) \mapsto (g, t - t_g + T_{j^*}),$$

is an isomorphism; this proves the second part of a).

Since $j^* > j_0$ and P acts uniserially on T, the group S/T_{j^*}, and hence also G/N_{j^*}, has coclass r. In particular, if S/T_{j^*} has order p^u, then $T_{j^*} = \gamma_{u-r+1}(S)$. The isomorphism α now implies that $\gamma_{u-r+1}(G) \leqslant N_{j^*}$. Note that G/N_{j^*} acts uniserially on N_{j^*}, and $\gamma_{u-r+1}(G)$ is G/N_{j^*}-invariant. Thus, G has coclass r if $\gamma_{u-r+1}(G) = N_{j^*}$, which holds if $\gamma_{u-r+1}(G)$ is not contained in N_{j^*+1}. We now show that this is the case.

Since $\gamma_{u-r}(S/T_{j^*}) = T_{j^*-1}/T_{j^*}$, we deduce that $\gamma_{u-r}(G/N_{j^*}) = N_{j^*-1}$ via α, so there exist $(1, t + T_m) \in \gamma_{u-r}(G)$ and $(g, -t_g + T_m) \in G$ such that $t \in T_{j^*-1}$ and $t - t^g \notin T_{j^*+1}$. But $j^* > j_0$, so $t \in T_{j^*-1} \leqslant pT$, and therefore $f(t \wedge T^*) \subseteq T_{j^*+1}$; a short calculation now confirms that the following holds in G modulo N_{j^*+1}:

$$(g, -t_g + T_{j^*+1})(1, t + T_{j^*+1}) = (1, t + T_{j^*+1})(g, -t_g + T_{j^*+1})(1, t - t^g + T_{j^*+1}).$$

Since $t - t^g \notin T_{j^*+1}$, this shows that $[(g, -t_g + T_{j^*+1}), (1, t + T_{j^*+1})]$ is in $\gamma_{u-r+1}(G)$ but not in N_{j^*+1}. Thus, G has coclass r. This proves a), and b) follows immediately.

To prove c), suppose $x = 0$, that is, $S = P \ltimes T$ splits and $G = G_{f,m}$ by Remark 8.16b). Since $f : T \wedge T \to T_j$ is surjective, the quotient of $T_{f,m}$ by T_{j+1}/T_m is not abelian. This shows S/T_{j+1} is not a quotient of $G_{f,m}$. Now a) proves the claim.□

Corollary 8.13 can now be generalised as follows.

Corollary 8.18 *Using the notation of Theorem 8.17, if there is a surjective $\mathbb{Z}_pP$-module homomorphism $T \wedge T \to T_j$, then the coclass tree associated with S has unbounded depth.*

8.4 The isomorphism problem for skeleton groups

We have seen that skeleton groups have an explicit construction in terms of homomorphisms. It remains to decide when different homomorphisms yield isomorphic groups: the isomorphism problem for skeleton groups.

We make the following assumptions throughout this section: S is an infinite pro-p-group of coclass r and dimension d, with translation subgroup T and corresponding point group P. As in Section 8.3.2, we denote by $S^* = P.T^* = P^* \ltimes T^*$ with $T^* = p^{-x}T$ the supergroup of S defined in Theorem 8.14. We assume that the translation subgroup T of S is characteristic in S. This holds, for example, if S is a space group or if the translation subgroup is chosen as $\gamma_\ell(S)$ for a suitable ℓ, see Corollary 4.40. In this case, every $\alpha \in \mathrm{Aut}(S)$ can naturally be extended to an automorphism of S^*.

Proposition 8.19 *Let $G_{P^*,f,m}$ and $G_{P^*,f',m}$ be skeleton groups, and let $\alpha \in \mathrm{Aut}(S)$ such that*

$$\alpha(f(s \wedge t)) \equiv f'(\alpha(s) \wedge \alpha(t)) \bmod T_{m+2xd} \quad \text{for all } s, t \in T. \tag{8.9}$$

a) *If $S = S^*$ is split, then $G_{P^*,f,m} \cong G_{P^*,f',m}$.*
b) *If S is non-split, then $G_{P^*,f,m} \cong G_{\alpha(P^*),f',m}$.*

Proof By assumption, α induces automorphisms of T and of $P \cong S/T$ which we denote by α_T and α_P, respectively. In particular, α maps $(g, t) \in P.T$ to $(\alpha_P(g), \phi(g) + \alpha_T(t))$ for some map $\phi\colon P \to T$.

a) If S is split, then $x = 0$, $P^* = P$, and $T^* = T$. Moreover,

$$S \to S, \quad (g, t) \mapsto (\alpha_P(g), \alpha_T(t)),$$

is an automorphism satisfying (8.9), so we can assume $\phi = 0$. Together with (8.9), a direct calculation shows that

$$P \ltimes T_{f,m} \to P \ltimes T_{f',m}, \quad (g, t + T_m) \mapsto (\alpha_P(g), \alpha_T(t) + T_m),$$

is a well-defined isomorphism between $G_{P^*,f,m}$ and $G_{P^*,f',m}$.

b) Since $T^* = p^{-x}T$, we can extend α_T to an automorphism of T^* by mapping $p^{-x}t$ with $t \in T$ to $p^{-x}\alpha_T(t)$. We denote the induced automorphism of S^* also by α; it maps $(g, t) \in P.T^*$ to $(\alpha_P(g), \phi(t) + \alpha_{T^*}(t))$. Since $\alpha(T^*) = T^*$, the group $\alpha(P^*)$ is a complement to T^* in S^*. Recall that $P^* = \{h_g \mid g \in S\}$ with each $h_g = (g, t_g) \in P.T^*$. Similarly, we write the elements of $\alpha(P^*)$ as $k_g = (g, s_g) \in P.T^*$ such that for $g \in P$ we have

$$k_{\alpha_P(g)} = \alpha(h_g) = (\alpha_P(g), \phi(g) + \alpha_{T^*}(t_g)) \in P.T^*.$$

The latter defines $s_{\alpha_P(g)}$, and it follows from $\phi(g) \in T$ that

$$(*) \quad s_{\alpha_P(g)} \equiv \alpha_{T^*}(t_g) \bmod T.$$

We now construct an isomorphism between the two skeleton groups

$$G_{P^*,f,m} = \{(h_g, t - t_g + T_m) \mid g \in P, t \in T\} \leqslant P^* \ltimes T^*_{f,m}$$
$$G_{\alpha(P^*),f,m} = \{(k_g, t - s_g + T_m) \mid g \in P, t \in T\} \leqslant \alpha(P^*) \ltimes T^*_{f',m}.$$

Since $T^* = p^{-x}T$, Condition (8.9) also holds for all $s, t \in T^*$ if 'mod T_{m+2xd}' is replaced by 'mod T_m'. A direct calculation using (8.9) shows that the following is an isomorphism

$$\psi\colon P^* \ltimes T^*_{f,m} \to \alpha(P^*) \ltimes T^*_{f',m}, \quad (h_g, t + T_m) \mapsto (k_{\alpha_P(g)}, \alpha_{T^*}(t) + T_m).$$

Now $G_{P^*,f,m}$ is a subgroup of $P^* \ltimes T^*_{f,m}$, and ψ maps $(h_g, t - t_g + T_m) \in G_{P^*,f,m}$ to $(k_{\alpha_P(g)}, \alpha_{T^*}(t) - \alpha_{T^*}(t_g) + T_m)$. By $(*)$, the latter is an element of $G_{\alpha(P^*),f',m}$, so ψ induces an isomorphism $G_{P^*,f,m} \cong G_{\alpha(P^*),f',m}$, as claimed. □

Variations of Proposition 8.19 are in the literature, for example, [33, Lemma 3.3]. Our proposition is a slightly amended version of that lemma.

Remark 8.20 Isomorphisms between skeleton groups arising from automorphisms of S (as described in Proposition 8.19) are labelled *orbit isomorphisms* in [33]. If there exists an orbit isomorphism between two isomorphic skeleton groups, then the skeleton subgraph is essentially determined by the structure of the associated pro-p-

group S, a favourable situation. We see in Section 8.4.2 that this situation occurs for the case of cyclic point groups. But other isomorphisms (*exceptional isomorphisms*) can exist, and it is possible that two homomorphisms define isomorphic skeleton groups for the split supergroup S^*, but the associated skeleton groups for S are non-isomorphic; we refer to [33, p. 1249 & 1269] for an example.

8.4.1 The split case

We now focus on the isomorphism problem for skeleton groups defined for split pro-p-groups. In the following, let $S = P \ltimes T$ be a pro-p-group of coclass r and dimension d with characteristic translation subgroup T. Recall that the twisted group $T_{f,m}$ and $T_{f',m}$ have the same underlying set T/T_m; this is used in part b) of the following lemma, where T/T_m denotes the usual quotient group of $(T, +)$ by $(T_m, +)$.

Lemma 8.21 *Let $j \leqslant m \leqslant 2j - d$ and consider surjective $\mathbb{Z}_p P$-homomorphisms $f, f' \colon T \wedge T \to T_j$.*

a) *Every automorphism of T/T_m lifts to an automorphism of T.*

b) *Every isomorphism $\beta \colon T_{f,m} \to T_{f',m}$ induces an automorphism of T/T_m.*

c) *There exists an isomorphism $T_{f,m} \to T_{f',m}$ if and only if there exists $\alpha \in \mathrm{Aut}(T)$ with $\alpha(T_m) = T_m$ such that (8.9) is satisfied for $x = 0$.*

Proof a) Let $\beta \in \mathrm{Aut}(T/T_m)$ and identify $\mathrm{Aut}(T)$ with $\mathrm{GL}_d(\mathbb{Z}_p)$ by fixing a $\mathbb{Z}_p$-basis $\{t_1, \ldots, t_d\}$ of T. Choose a $d{\times}d$ matrix $A = (a_{i,j})$ over $\mathbb{Z}_p$ such that $\beta(t_j{+}T_m) = a_{j,1}t_1 + \ldots + a_{j,d}t_d + T_m$ for each j. If $m \geqslant d$, then $T_m \leqslant pT \leqslant T$ and so A mod p describes the restriction of β to T/pT, which is an automorphism of T/pT. This implies p does not divide $\det(A)$, and so $\det(A)$ is a unit in $\mathbb{Z}_p$. It follows that $A \in \mathrm{GL}_d(\mathbb{Z}_p)$ is invertible. (This standard result can be found, for example, in [21, Theorem 11.30].) By our identification, A is an automorphism of T, and, by construction, it induces β. Thus, β lifts to an automorphism of T, as claimed. The case $m < d$ follows similarly: now T/T_m is elementary abelian and we can choose a subset of $\{t_1 + T_m, \ldots, t_d + T_m\}$ as basis.

b) Note that $T_{f,m}$, $T_{f',m}$, and T/T_m have the same underlying set. Let $\alpha \colon T \to T$ be a map such that $\beta(t + T_m) = \alpha(t) + T_m$ for all $t \in T$. Since β is an isomorphism and $[s + T_m, t + T_m] = f(s \wedge t) + T_m$ in $T_{f,m}$ for all $s, t \in T$, we deduce that $\alpha(f(s \wedge t)) \equiv f'(\alpha(s) \wedge \alpha(t)) \bmod T_m$. Thus,

$$(*) \quad \beta(s + T_m) \circ_{f'} \beta(t + T_m) = \alpha(s) + \alpha(t) + \tfrac{1}{2}\alpha(f(s \wedge t)) + T_m.$$

On the other hand, we can write $(s + T_m) \circ_f (t + T_m) = (x + T_m) \circ_f (y + T_m)$ where $x = s + t$ and $y = \frac{1}{2}f(s \wedge t)$ with $f(x \wedge y) \in T_m$. This yields

$$\begin{aligned}\beta((s + T_m) \circ_f (t + T_m)) &= \alpha(x) + \alpha(y) + \tfrac{1}{2}f'(\alpha(x) \wedge \alpha(y)) + T_m \\ &= \alpha(x) + \alpha(y) + T_m + \tfrac{1}{2}\alpha(f(x \wedge y)) + T_m \\ &= \alpha(s + t) + \tfrac{1}{2}\alpha(f(s \wedge t)) + T_m.\end{aligned}$$

Together with $(*)$, we deduce that $\alpha(s+t)+T_m = \alpha(s)+\alpha(t)+T_m$ for all $s,t \in T$. Thus, β induces an automorphism $t+T_m \mapsto \alpha(t)+T_m$ of T/T_m, as claimed.

c) If $\beta\colon T_{f,m} \to T_{f',m}$ is an isomorphism, then $\beta \in \mathrm{Aut}(T/T_m)$ by b), so $\beta(t+T_m) = \alpha(t)+T_m$ for some $\alpha \in \mathrm{Aut}(T)$ with $\alpha(T_m) = T_m$ by a); one can check that α satisfies (8.9) with $x = 0$. Conversely, a direct calculation shows that every α as in the lemma yields an isomorphism $T_{f,m} \to T_{f',m}$, $t+T_m \mapsto \alpha(t)+T_m$. □

The following proposition gives a partial converse of Proposition 8.19a).

Proposition 8.22 *Let $G_{f,m}$ and $G_{f',m}$ be skeleton groups where $f, f'\colon T \wedge T \to T_j$ are surjective and $j \leqslant m \leqslant 2j-d$. If there is an isomorphism $\phi\colon G_{f,m} \to G_{f',m}$ with $\phi(T_{f,m}) = T_{f',m}$, then the following hold.*

a) *The isomorphism ϕ induces an automorphism β of $S/T_j = P \ltimes T/T_j$.*

b) *If $j > d$ and the automorphism β in* a) *lifts to an automorphism $\alpha \in \mathrm{Aut}(S)$, then α satisfies* (8.9) *with $x = 0$, and so there is an orbit isomorphism $G_{f,m} \to G_{f',m}$.*

Proof a) Lemma 8.21 and $\phi(T_{f,m}) = T_{f',m}$ imply that there exist an automorphism $\psi \in \mathrm{Aut}(T)$ and a map $\tau\colon P \to T$ such that ϕ maps $(g, t+T_m) \in G_{f,m}$ to $(\phi_P(g), \tau(g)+\psi(t)+T_m)$. Since ϕ is an isomorphism,

$$(\dagger) \quad \psi(f(s \wedge t)) \equiv f'(\psi(s) \wedge \psi(t)) \bmod T_m \quad \text{for all } s,t \in T.$$

Moreover, ϕ maps the derived subgroup T_j/T_m of $T_{f,m}$ to that of $T_{f',m}$, so ψ induces an automorphism of T/T_j. Since f has image T_j, the homomorphism property of ϕ implies that $\psi(t^g) \equiv \psi(t)^{\phi_P(g)} \bmod T_j$ and $\tau(gh) \equiv \tau(g)^{\phi_P(h)} + \tau(h) \bmod T_j$ for all $t \in T$ and $g, h \in P$. It follows that ϕ induces the automorphism

$$\beta\colon S/T_j \to S/T_j, \quad (g, t+T_j) \mapsto (\phi_P(g), \tau(g)+\psi(t)+T_j).$$

In fact, also $\beta'\colon (g, t+T_j) \mapsto (\phi_P(g), \psi(t)+T_j)$ defines an automorphism of S/T_j.

b) Let β be as in the proof of a). If β lifts to an automorphism $\alpha \in \mathrm{Aut}(S)$, then α has the form $(g,t) \mapsto (\phi_P(g), \tau'(g)+\alpha_T(t))$ for some map $\tau'\colon P \to T$ such that $\tau'(g) \equiv \tau(g) \bmod T_j$ for all $g \in P$ and

$$(*) \quad \alpha_T(t) \equiv \psi(t) \bmod T_j \quad \text{for all } t \in T.$$

Recall that $d \leqslant j \leqslant m \leqslant 2j-d$, and write $j = ad+b$ with $b \in \{0,\ldots,d-1\}$ and $a \geqslant 1$. Then $j \geqslant ad$ and $j+ad = 2j-b \geqslant 2j-d \geqslant m$, that is, $T_j \leqslant p^aT$ and $p^aT_j \leqslant T_m$. Together with $(*)$, this implies that

$$(\ddagger) \quad \alpha_T(s) \equiv \psi(s) \bmod T_m \quad \text{for all } s \in T_j.$$

Since f has image T_j and $f'(T \wedge T_m) \leqslant T_m$, it follows that for every $s,t \in T$

$$\alpha_T(f(s \wedge t)) \overset{(\ddagger)}{\equiv} \psi(f(s \wedge t)) \overset{(\dagger)}{\equiv} f'(\psi(s) \wedge \psi(t)) \overset{(\ddagger)}{\equiv} f'(\alpha_T(s) \wedge \alpha_T(t)) \bmod T_m.$$

This shows that α satisfies (8.9) with $x = 0$, and Proposition 8.19 shows that there is an orbit isomorphism between $G_{f,m}$ and $G_{f',m}$, as claimed. The same argument proves that there is an orbit isomorphism if β' lifts to an automorphism of S. □

8.4.2 Cyclic point groups: orbit isomorphisms

We now consider uniserial p-adic space groups S with cyclic point group, and show that all isomorphisms between skeleton groups defined by S are orbit isomorphisms, that is, induced by automorphisms of S; this result first appeared in [18]. Recall from Proposition 4.32 that the translation subgroup of S is unique, and therefore characteristic. We start with a preliminary lemma for split groups.

Lemma 8.23 *Let $S = P \ltimes T$ be a uniserial p-adic space group of dimension d with P of nilpotency class c and $\gamma_{c+1}(S) = T_k$. Let $G_{f,m}$ and $G_{f',m}$ be skeleton groups where $f, f' \colon T \wedge T \to T_j$ are onto with $j > \max\{k, d+1\}$. If $C_P(T_k/T_j) = \{1\}$, then any isomorphism $G_{f,m} \to G_{f',m}$ maps $T_{f,m}$ to $T_{f',m}$.*

Proof Recall that $G = G_{f,m}$ has underlying set $P \times T/T_m$ and a subgroup $N = T_j/T_m$. The proof of Theorem 8.17 shows that $S/T_j \cong G/N$ and, if $|S/T_j| = p^u$, then $\gamma_{u-r+1}(G) = N$, where r is the coclass of S. Moreover, $T_k/T_j = \gamma_{c+1}(S/T_j)$ is isomorphic to $\gamma_{c+1}(G/N)$, so $\gamma_{c+1}(G) = T_k/T_m$. Let C be the centraliser in G of $\gamma_{c+1}(G)/\gamma_{u-r+1}(G) \cong T_k/T_j$. Clearly, $T_{f,m} \leqslant C$ and, by assumption, $C \cap P = \{1\}$. It follows that $C = T_{f,m}$ and, analogously, $C = T_{f',m}$. This implies the claim. □

Proposition 8.24 *Let $S = P.T$ be a uniserial p-adic space group with cyclic point group P. Let $G_{P^*,f,m}$ and $G_{P^*,f',m}$ be skeleton groups where $f, f' \colon T \wedge T \to T_j$ are onto. If $j > d$ and $G_{P^*,f,m} \cong G_{P^*,f',m}$, then there is an orbit isomorphism $G_{P^*,f,m} \to G_{P^*,f',m}$.*

Proof Proposition 5.33 shows that for every $r \geqslant 1$ there is, up to isomorphism, a unique uniserial p-adic space group with cyclic point group of order p^r, namely the split group $S = S(p^r)$ with translation subgroup $T = (\mathbb{Z}_p[\theta], +)$ and point group $P = \langle \theta \rangle$ for a primitive p^r-th root of unity θ. The uniserial series of T has terms $T_i = (\theta - 1)^i T$ for all i, and each T_i is characteristic in S. In particular, $T_i = \gamma_{i+1}(S)$ for each $i \geqslant 1$.

Let $\phi \colon G_{f,m} \to G_{f',m}$ be an isomorpism. If $\theta^{(p^i)} \in P$ centralises T_1/T_j, then $(\theta^{(p^i)} - 1)t \in T_j$ for all $t \in T_1$, so $\theta^{(p^i)} - 1 \in T_{j-1}$. Since $T_j > pT$, this forces $\theta^{(p^i)} = 1$, and therefore $C_P(T_1/T_j) = \{1\}$. Lemma 8.23 shows that $\phi(T_{f,m}) = T_{f',m}$. This, in turn, allows us to apply Proposition 8.22, which proves that ϕ induces an automorphism β of $S/T_j = P \ltimes T/T_j$. In fact, the proof of Proposition 8.22a) shows that there is an automorphism of the form

$$\beta' \colon (\theta, t + T_j) \mapsto (\phi_P(\theta), \phi_{T/T_j}(t + T_j)),$$

with $\phi_P(\theta) = \theta^k$ for some k. We now show that β' lifts to an automorphism of S. Note that $t^{\phi_P(\theta)} = t^{(\theta^k)} = \theta^k t = \sigma_k(\theta)t$ for all $t \in T$, where σ_k is the automorphism of T defined by $\theta \mapsto \theta^k$. Let $v_0 \in T$ such that $\phi_{T/T_j}(1+T_j) = v_0 + T_j$; in fact, v_0 is a unit in $\mathbb{Z}_p[\theta]$, since otherwise ϕ_{T/T_j} would not be an automorphism of T/T_j. Together, if $i \geqslant 1$, then

$$\phi_{T/T_j}(\theta^i + T_j) = \phi(1^{(\theta^i)} + T_j) = \phi(1+T_j)^{\phi(\theta^i)} = v_0\sigma_k(\theta^i) + T_j,$$

so $\phi_{T/T_j}(t+T_j) = v_0\sigma_k(t) + T_j$ for all $t \in T$. Define $\psi \in \mathrm{Aut}(T)$ by $t \mapsto v_0\sigma_k(t)$. By construction, $\psi(t^\theta) = \psi(t)^{\phi_P(\theta)}$, so $\alpha\colon S \to S$, $(\theta, t) \mapsto (\phi_P(\theta), \psi(t))$, is an automorphism that induces β'. The proof of Proposition 8.22b) now shows that there is an orbit isomorphism between $G_{f,m}$ and $G_{f',m}$, as claimed. □

We conclude this section with a result on Periodicity I for skeletons. Let $S = P \ltimes T$ be a pro-p-group of coclass r and dimension d with characteristic translation subgroup T. Recall that the j-th skeleton $\mathcal{S}_j$ is the subtree of the j-th branch $\mathcal{B}_j$ that is generated by all skeleton groups $G_{f,m} = P \ltimes T_{f,m}$; this requires that each $f\colon T \wedge T \to T_j$ is a surjective $\mathbb{Z}_pP$-homomorphism and $j \leqslant m \leqslant 2j - d$. If such an f exists, then the j-th skeleton is a rooted tree of depth $j - d$, see Theorem 8.17. Multiplication by p yields a $\mathbb{Z}_pP$-homomorphism $pf\colon T \wedge T \to T_{j+d}$ that can be used to define the skeleton group $G_{pf,m+d}$ in $\mathcal{S}_{j+d}$. Suppose now that there always exists an orbit isomorphism between every pair of isomorphic skeleton groups in $\mathcal{S}_j$; as shown above, this holds, for example, for uniserial space groups with cyclic point group. In this case, mapping $G_{f,m}$ to $G_{pf,m+d}$ induces a graph isomorphism between the skeleton $\mathcal{S}_j$ and the pruned skeleton $\mathcal{S}_{j+d}(j-d)$; recall the definition of a pruned tree from Definition 7.1. We summarise this result in the following proposition.

Proposition 8.25 *Let $S = P \ltimes T$ be a pro-p-group of coclass r and dimension d with characteristic translation subgroup T. Suppose there exists a surjective $\mathbb{Z}_pP$-homomorphism $T \wedge T \to T_j$ with $j > d$, and suppose every pair of isomorphic skeleton groups in $\mathcal{S}_j$ and in $\mathcal{S}_{j+d}$ admits an orbit isomorphism. Then the skeleton $\mathcal{S}_j$ is isomorphic as a rooted tree to the pruned skeleton $\mathcal{S}_{j+d}(j-d)$.*

Proof Multiplication by p induces a bijection between the surjective $\mathbb{Z}_pP$-homomorphisms $T \wedge T \to T_j$ and $T \wedge T \to T_{j+d}$. Moreover, if $j \leqslant m \leqslant 2j - d$, then $j + d \leqslant m + d \leqslant 2j$. Let $f, f'\colon T \wedge T \to T_j$ be surjective. By assumption, $G_{f,m}$ and $G_{f',m}$ are isomorphic if and only if there exists some $\alpha \in \mathrm{Aut}(S)$ that satisfies (8.9) with $x = 0$; analogous for $G_{pf,m+d}$ and $G_{pf',m+d}$. Since (8.9) holds for f and f' and m if and only if it holds for pf and pf' and $m+d$, we deduce that $G_{f,m} \cong G_{f',m}$ if and only if $G_{pf,m+d} \cong G_{pf',m+d}$. Lastly, Theorem 8.17 shows that the immediate ancestor of $G_{f,m}$ is $G_{f,m-1}$, and the immediate ancestor of $G_{pf,m+d}$ is $G_{pf,m-1+d}$. The claim follows. □

8.5 Constructible groups versus skeleton groups

The aim of this section is to reconcile our definition of skeleton groups with the constructible groups defined in [57, Definition 8.4.9]. The original definition of constructible groups is only for uniserial p-adic space groups.

Proposition 8.26 *Let $p > 2$ be a prime. Every constructible group (as defined in [57, Definition 8.4.9]) is a skeleton group, and every skeleton group for a uniserial p-adic space group is a constructible group.*

Proof Let $S = P.T$ be a uniserial p-adic space group of coclass r and dimension d, with point group P, translation subgroup T, and extended uniserial series with terms T_i for $i \in \mathbb{Z}$. We now follow [57, Section 8.4] and describe the construction of constructible groups; going through the details of this construction will reveal that they are skeleton groups as defined in Definition 8.15. This construction involves *pull-backs* and *Baer sums* of group extensions; while we provide cohomological interpretations of these constructions, we refer to [57, Section 2.2] for the precise definitions and discussions.

Constructible groups. Let S^* be the split supergroup of S defined in Theorem 8.14, that is, $S^* = P.T^* = P^* \ltimes T^*$ with $T^* = p^{-x}T$ and embedding $\iota\colon S \to S^*$. To keep the notation simple, by slight abuse of notation, we also denote the induced embeddings $S/T_i \to S^*/T_i$ by ι for $i \in \mathbb{N}$. Let $f'\colon T \wedge T \to T_j$ be a surjective $\mathbb{Z}_pP$-homomorphism, inducing a surjective $f\colon (T^*/T_{j^*}) \wedge (T^*/T_{j^*}) \to T_{j^*}/T_m$ where we write $j^* = j - 2xd$ as before; this requires $f'(T_{j^*} \wedge T^*) \leqslant T_m$. Now define

$$R(f) = (T^*/T_{j^*} \times T_{j^*}/T_m, \circ_f) \tag{8.10}$$

with multiplication $(a, c) \circ_f (b, d) = (a + b, c + d + \frac{1}{2} f(a \wedge b))$. A short calculation shows that $R(f)$ is a group extension of T^*/T_{j^*} by T_{j^*}/T_m.

Below we represent group extensions by short exact sequences; with this convention, $R(f)$ is given by the short exact sequence

$$\mathcal{R}_f \quad \equiv \quad 0 \to T_{j^*}/T_m \to R(f) \to T^*/T_{j^*} \to 1$$

where the embedding and projection maps are the usual ones; we use '$\equiv$' do denote equivalent extensions, see Proposition 5.9. The extension $R(f)$ has the same commutator structure as the group $T^*_{f',m}$ defined in Definition 8.11. In fact, if

$$\mathcal{E} \quad \equiv \quad 0 \to T_{j^*}/T_m \to T^*/T_m \to T^*/T_{j^*} \to 1$$

describes T^*/T_m as an extension of T^*/T_{j^*} by T_{j^*}/T_m, then $T^*_{f',m}$ can be defined by the Baer sum $\mathcal{R}_f + \mathcal{E}$: in cohomological terms, if δ and γ are cohomology class describing the extensions $R(f)$ and T^*/T_m, respectively, then the extension specified by the Baer sum $\mathcal{R}_f + \mathcal{E}$ is described by $\delta + \gamma$.

Since f is compatible with the action of P, one can define the semidirect product $P^* \ltimes R(f)$; let this extension of S^*/T_{j^*} by T_{j^*}/T_m be represented by a short exact sequence $\mathcal{F}_f$. Now one can use ι and a pull-back construction to define a group Y via the short exact sequence $\mathcal{F}_f\,\iota$; specifically, there is a group Y and homomorphism $\iota'\colon Y \to P^* \ltimes R(f)$ such that the following diagram of short exact sequences in commutative:

$$\begin{array}{lcccccccc}
\mathcal{F}_f & \equiv 0 \longrightarrow & T_{j^*}/T_m & \longrightarrow & P^* \ltimes R(f) & \longrightarrow & S^*/T_{j^*} & \longrightarrow & 1 \\
& & \Vert & & \uparrow{\scriptstyle \iota'} & & \uparrow{\scriptstyle \iota} & & \\
\mathcal{F}_f\,\iota & \equiv 0 \longrightarrow & T_{j^*}/T_m & \longrightarrow & Y & \longrightarrow & S/T_{j^*} & \longrightarrow & 1.
\end{array}$$

In cohomological terms, if δ is the cohomology class defining $P^* \ltimes R(f)$ as an extension of S^*/T_{j^*} by T_{j^*}/T_m, then the restriction of δ to $\iota(S/T_{j^*})$ describes $\iota'(Y)$ as an extension of $\iota(S/T_{j^*})$ by T_{j^*}/T_m; now Y is the preimage of that extension under ι'. For the last step, we need to describe S/T_m as an extension of S/T_{j^*}, say

$$\mathcal{D} \;\equiv\; 0 \to T_{j^*}/T_m \to S/T_m \to S/T_{j^*} \to 1.$$

Finally, the *constructible group* G_f is defined via the Baer sum

$$\mathcal{D} + \mathcal{F}_f\,\iota \;\equiv\; 0 \to T_{j^*}/T_m \to G_f \to S/T_{j^*} \to 1.$$

This completes the description of constructible groups as in [57]; we now investigate this further and explain how the construction is related to skeleton groups. We start with the case that S splits over T.

Split case. If S splits over T, then the construction of $S^* = P^* \ltimes T = P.T = S$ amounts to choosing a complement P^*. The embedding $\iota\colon S \to S^*$ is an isomorphism, and working through the above details shows that G_f is isomorphic to the skeleton group $G_{f,m} = P \ltimes T_{f,m}$, as claimed. In particular, the isomorphism type of G_f is independent of the chosen complement and embedding; starting with $S = P \ltimes T$, we can assume that ι is the identity, so $G_{f,m}$ can be defined via the Baer sum $\mathcal{D} + \mathcal{F}_f$.

Non-split case. Now suppose that S does not split over T. To understand the group defined by $\mathcal{D} + \mathcal{F}_f\,\iota$, we detour via $S^* = P^* \ltimes T^*$. As explained in the previous paragraph, G^*_f can be constructed as $G^*_{f,m} = P^* \ltimes T^*_{f,m}$, which is realised via the Baer sum $\tilde{\mathcal{D}} + \mathcal{F}_f$, where

$$\tilde{\mathcal{D}} \;\equiv\; 0 \to T_{j^*}/T_m \to S^*/T_m \to S^*/T_{j^*} \to 1.$$

Here we identify $S^*/T_i = P^* \ltimes T^*/T_i$ for $i \geqslant -xd$. A short calculation shows that the extensions $\mathcal{D}$ and $\tilde{\mathcal{D}}\iota$ are equivalent, and now it follows from [57, Exercise 2.2(3)(iii)] that

$$\mathcal{D} + \mathcal{F}_f\,\iota \;\equiv\; \tilde{\mathcal{D}}\iota + \mathcal{F}_f\,\iota \;\equiv\; (\tilde{\mathcal{D}} + \mathcal{F}_f)\iota.$$

Thus, $\mathcal{D}+\mathcal{F}_f\,\iota$ can be constructed via a pull-back as follows; recall from the discussion of the split case that we can assume G^*_f is constructed as $G^*_{f,m} = P^* \ltimes T^*_{f,m}$:

$$\begin{array}{lcccccccc}
\tilde{\mathcal{D}} + \mathcal{F}_f & \equiv 0 \longrightarrow & T_{j^*}/T_m & \longrightarrow & G^*_{f,m} & \overset{\pi}{\longrightarrow} & S^*/T_{j^*} & \longrightarrow & 1 \\
& & \Vert & & \uparrow{\scriptstyle \iota''} & & \uparrow{\scriptstyle \iota} & & \\
(\tilde{\mathcal{D}} + \mathcal{F}_f)\iota & \equiv 0 \longrightarrow & T_{j^*}/T_m & \longrightarrow & G_f & \longrightarrow & S/T_{j^*} & \longrightarrow & 1.
\end{array}$$

In cohomological terms, if δ is the cohomology class defining $G^*_{f,m}$ as an extension of S^*/T_{j^*} by T_{j^*}/T_m, then the restriction of δ to $\iota(S/T_{j^*})$ describes $\iota''(G_f)$ as an extension of $\iota(S/T_{j^*})$ by T_{j^*}/T_m; now G_f is the preimage of that extension under the embedding ι''. In other words, $\iota''(G_f) \cong G_f$ is the full preimage of $\iota(S/T_{j^*})$ under π. Comparing this with the construction in Section 8.3.2, we see that G_f is a skeleton group, as claimed. It follows that every skeleton group for a space group is constructible. □

Proposition 8.26 considers odd primes. By [57, Definition 8.4.9], a finite 2-group of coclass r is constructible if it is isomorphic to S/U where S is an infinite uniserial 2-adic space group of coclass at most r and U is a sublattice of the translation subgroup of S.

In [57] constructible groups are only defined for uniserial p-adic space groups, and we saw in Proposition 8.26 that in this case constructible groups are exactly the skeleton groups. What happens for infinite pro-p-groups that are not space groups? If $G_{P^*f,m}$ is a skeleton group for an infinite pro-p-group S with hypercentre H, then Corollary 4.40 implies that $G_{P^*,f,m}$ has a normal subgroup $M \cong H$ such that $G_{P^*,f,m}/M$ is isomorphic to a skeleton group for the uniserial p-adic space group S/H; see also [57, Lemma 8.4.10]. This allows us to state [57, Theorems 11.3.7 and 11.3.9] for skeleton groups as follows.

Theorem 8.27 ([57, Theorems 11.3.7 and 11.3.9]) *There is a constant $s = s(p,r)$ such that almost every finite p-group G of coclass r has a normal subgroup N such that G/N is a skeleton group and $|N| \leqslant s(p,r)$; if p is odd, then $s(p,r) = (p-1)p^{r-1}(7p^r + 7r + 4)$, and $s(2,r) = 2^{r-1}(2^{r-1} + r + 3)$ otherwise.*

An immediate corollary for $p = 2$ is the following.

Corollary 8.28 *The branches of the coclass trees in $\mathcal{G}(2,r)$ have depth at most $2^{r-1}(2^{r-1}+r+3)$. Therefore every coclass graph $\mathcal{G}(2,r)$ can be completely described by a finite subgraph and the periodic pattern defined by Periodicity I.*

Chapter 9
Groups of maximal class

In this chapter, we explore the coclass graphs $\mathcal{G}(p, 1)$ for $p \geqslant 5$ in greater depth, with a particular focus on a possible Periodicity II. More precisely, we consider the branches of the unique maximal coclass tree in $\mathcal{G}(p, 1)$ and for each branch we introduce a partition of its skeleton into *Galois trees*. The main result of this chapter is a complete description of the Galois trees associated with the maximal *Galois order*. In the final section of this chapter we outline how this could provide a way towards proving a Periodicity II.

The structure of $\mathcal{G}(5, 1)$ is experimentally well understood, see [17, 66], and significant parts of this structure are proved, see [13,14]. For $p \geqslant 7$, an experimental description is difficult to obtain; one reason for this is that the branches of the coclass tree in $\mathcal{G}(p, 1)$ have unbounded depth and width, and hence are far too 'thick' for an explicit visualisation. The results of this chapter provide evidence for this difficulty.

9.1 The broad structure of $\mathcal{G}(p, 1)$

We first recall a well-known invariant of finite p-groups.

Definition 9.1 Let G be a finite p-group of coclass 1 and order p^n with $n \geqslant 4$. For $i \geqslant 2$ write $P_i(G) = \gamma_i(G)$ and define $P_1(G) = C_G(P_2(G)/P_4(G))$. The *degree of commutativity* $\mathrm{doc}(G)$ of G is the largest integer ℓ such that

$$[P_i(G), P_j(G)] \leqslant P_{i+j+\ell}(G)$$

for all i, j, and $\ell = n - 2$ if $P_1(G)$ is abelian.

The following proposition summarises results of Blackburn [1], Leedham-Green & McKay [57, Section 3.2], and Fernández Alcober [38]. The proof is technical and we refer to [1, 38, 57] for details. Further results on the degree of commutativity for special cases have been obtained by Vera-López et al. [81].

H. Dietrich and B. Eick, *Coclass graphs of p-groups*, Springer Monographs in Mathematics, https://doi.org/10.1007/978-3-032-17506-9_9

Proposition 9.2 *Let G be a p-group of coclass 1 with order p^n and degree of commutativity ℓ. If $p = 2$, then $\ell = n - 2$; if $p = 3$, then $\ell \geqslant n - 4$; if $p = 5$, then $2\ell \geqslant n - 6$; if $p \geqslant 7$, then $2\ell \geqslant n - 2p + 5$.*

The degree of commutativity influences the broad structure of $\mathcal{G}(p, 1)$, as shown by the following theorem.

Theorem 9.3 *The general structure of $\mathcal{G}(p, 1)$ is visualised in Figure 9.1.*

a) *The graph $\mathcal{G}(p, 1)$ consists of the isolated point C_{p^2} and the maximal coclass tree $\Delta(C_p^2)$.*

b) *Let $S = S(p)$ be the pro-p-group of coclass 1 as in Proposition 3.16 and let $S_n = S/\gamma_n(S)$ for $n \in \mathbb{N}$. The mainline of $\Delta(C_p^2)$ is $S_2 \to S_3 \to \ldots$.*

c) *The depth of the branch $\mathcal{B}_n$ with root S_n is at most 1 if $p \in \{2, 3\}$, at most n if $p = 5$, and at most $n + 2p - 11$ if $p \geqslant 7$.*

Proof For $p \in \{2, 3\}$ this follows from Blackburn's classification, so let $p \geqslant 5$.

a) By Lemma 1.12, a group of coclass 1 has rank 1 (and is cyclic of order 4), or rank 2 and $G/\gamma_2(G) \cong C_2^2$ (and lies in $\Delta(C_2^2)$).

b) Let G be a group in $\mathcal{G}(p, 1)$ of order p^n; throughout, we write $P_i = P_i(G)$. Suppose P_1 is non-abelian, so $[P_1, P_1] = [P_1, P_2] \neq P_n = \{1\}$, and therefore $\mathrm{doc}(G) \leqslant n - 4$. If H is a descendant of G in $\mathcal{G}(p, 1)$ of order p^{n+e}, then $P_1(H)$ is non-abelian as well and $\mathrm{doc}(H) \leqslant \mathrm{doc}(G)$. Together with Proposition 9.2, for $p \geqslant 7$

$$n + e - 2p + 5 \leqslant 2\mathrm{doc}(H) \leqslant 2\mathrm{doc}(G) \leqslant 2n - 8.$$

Thus, if $p \geqslant 7$ and P_1 is non-abelian, then the depth of descendant tree $\Delta(G)$ is at most $n + 2p - 13$; for $p = 5$, the analogous calculation with $2\mathrm{doc}(H) \geqslant n + e - 6$ yields the bound $n - 2$. Hence, if G is a group on an infinite path, then P_1 necessarily is abelian.

Suppose $\mathrm{doc}(G) > 0$. We show that if $s \in G \setminus P_1$, then $s^p \in P_{n-1}$. Suppose, for a contradiction, that $s^p \in P_j \setminus P_{j+1}$ with $2 \leqslant j \leqslant n - 2$, and write $K = C_G(P_j/P_{j+2})$. Then P_j/P_{j+2} can be generated by $s^p P_{j+2}$ and $t P_{j+2}$ for some $t \in P_{j+1}$, and since s acts trivially on $P_{j+1}/P_{j+2} \cong C_p$, it follows that $s \in K$. Our assumption $\mathrm{doc}(G) > 0$ implies $P_1 \leqslant K$, and since $[G : P_1] = p$ and $K < G$, we conclude that $P_1 = K$. This is a contradiction to $s \notin P_1$, and so $s^p \in P_{n-1}$, as claimed. Note that this also implies that $s^p = 1$ for every group G that is not terminal in $\mathcal{G}(p, 1)$.

Now let G be a group on an infinite path in $\mathcal{G}(p, 1)$. We can assume that n is large enough so that $\mathrm{doc}(G) > 0$. The previous paragraphs show that $G = \langle s \rangle \ltimes P_1$ where $s \in G \setminus P_1$ has order p and P_1 is abelian. Choose $s_1 \in P_1 \setminus P_2$ and iteratively define $s_{i+1} = [s_i, s]$ for $i \geqslant 1$. By construction, $s_i \in P_i \setminus P_{i+1}$ for each $i \in \{1, \ldots, n - 1\}$.

Let $S = S(p) = P \ltimes T$ as in Proposition 3.16, that is, $P = \langle \theta \rangle \cong C_p$ and $T = (\mathbb{Z}[\theta], +)$ for a primitive p-th root of unity θ. Note that T is generated freely by $\{1, \theta, \theta^2, \ldots\}$ subject to the relations induced by $(\theta^p - 1)/(\theta - 1) = 0$. Writing $\kappa = \theta - 1$, it follows that T is freely generated by $\{1, \kappa, \kappa^2, \ldots\}$ subject to the relations induced by $((\kappa + 1)^p - 1)/\kappa = 0$, that is, $r_j = 0$ for $j \in \mathbb{N}_0$, where

$$r_j = \binom{p}{1}\kappa^j + \binom{p}{2}\kappa^{j+1} + \ldots + \binom{p}{p}\kappa^{j+d}.$$

To relate G with S, we consider the map $\sigma\colon T \to P_1$ that takes $t \in T$ written as $t = a_0 + a_1\kappa + \ldots + a_m\kappa^m$ with each $a_i \in \mathbb{Z}_p$ to

$$\sigma(t) = s_1^{a_0} s_2^{a_1} \cdots s_{m+1}^{a_m}.$$

First, we show that σ is well-defined. Since P_1 is abelian and

$$(*) \quad s_i^s = s_i[s_i, s] = s_i s_{i+1} \quad \text{for all } i,$$

we deduce that

$$(s s_{j+1})^p = s^p s_{j+1}^{\binom{p}{1}} s_{j+2}^{\binom{p}{2}} \cdots s_{j+p}^{\binom{p}{p}} \quad \text{for all } j.$$

Also $s, s s_{j+1} \in G \backslash P_1$ and $\mathrm{doc}(G) > 0$, so $s^p = (s s_{j+1})^p = 1$ for all $j \in \mathbb{N}_0$ as shown above. But then $\sigma(r_j) = (s s_{j+1})^p = 1$ for all j, which proves that σ is well-defined. By construction, σ is a surjective homomorphism. Since $\{s, s_1, \ldots, s_{n-1}\}$ is a polycyclic generating set for G and $s_j = 1$ for $j \geqslant n$, it follows that $t \in T$ lies in $\ker\sigma$ if and only if t is divisible by κ^{n-1}. Thus, $\ker\sigma = T_{n-1}$ is the ideal of T generated by κ^{n-1}, which corresponds to the subgroup $\gamma_n(S)$ of S.
Let $\theta \in P$ act on P_1 by conjugation by s. If $t = a_0 + a_1\kappa + \ldots + a_m\kappa^m \in T$, then $t\theta = t + \kappa t$, and so

$$\begin{aligned}\sigma(t\theta) &= \sigma(t + \kappa t) = (s_1 s_2)^{a_0}(s_2 s_3)^{a_1} \cdots (s_{m+1} s_{m+2})^{a_m} \\ &= (s_1^{a_0} s_2^{a_1} \cdots s_{m+1}^{a_m})^s = \sigma(t)^s = \sigma(t)^\theta.\end{aligned}$$

This proves that σ induces a P-isomorphism between $T/\gamma_n(S)$ and P_1, and so an isomorphism between $S = \langle\theta\rangle \ltimes T/T_{n-1}$ and $\langle\theta\rangle \ltimes P_1 \cong G$.
c) If G is an immediate descendant of S_n in $\mathcal{B}_n$, then $|G| = p^{n+1}$ and $P_1(G)$ is non-abelian. The first part of the proof of b) shows that $\Delta(G)$ has depth at most $(n+1)+2p-13$ if $p \geqslant 7$, and depth at most $(n+1)-2$ if $p = 5$. Thus $\mathcal{B}_n$ has depth at most $n + 2p - 11$ if $p \geqslant 7$, and depth at most n if $p = 5$. □

9.2 Skeleton groups in $\mathcal{G}(p,1)$

As before, let S be the infinite pro-p-group of coclass 1 and recall that $S = P \ltimes T$, where $P = \langle\theta\rangle$ and $T = (\mathbb{Z}_p[\theta], +)$ with $d = p - 1$. The branches in the coclass tree of $\mathcal{G}(p,1)$ have bounded depth for $p \in \{2,3\}$, unbounded depth but bounded width for $p = 5$, and unbounded depth and unbounded width for $p \geqslant 7$. These results essentially follow from Corollaries 8.28 and 8.13 and an investigation of the skeleton groups in $\mathcal{G}(p,1)$. Throughout the chapter, we assume that

$$p \geqslant 5, \quad d = p - 1, \quad \text{and} \quad \ell = (p-3)/2.$$

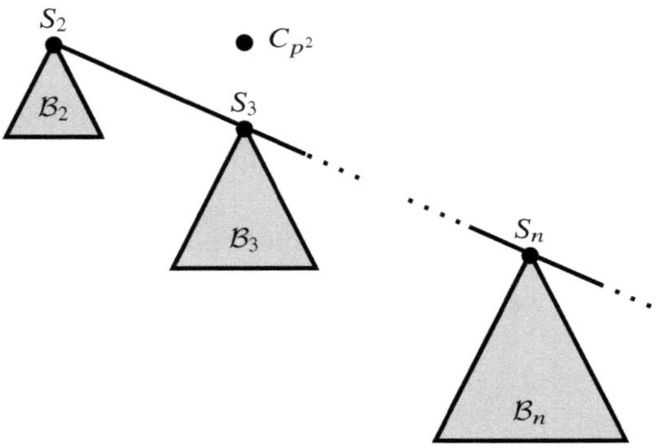

Fig. 9.1 The graph $\mathcal{G}(p, 1)$ has one isolated group and one maximal coclass tree with branches $\mathcal{B}_2, \mathcal{B}_3, \ldots$; the root of the branch $\mathcal{B}_n$ is $S_n = S/\gamma_n(S)$ of order p^n, where $S = S(p)$ is the infinite pro-p-group of coclass 1 in Proposition 3.16.

Many proofs in this section are from [6, 15] and rely significantly on the works by Leedham-Green & McKay cited below.

9.2.1 Homomorphisms

Proposition 8.10 shows that the $\mathbb{Z}_pP$-module $T \wedge T$ can be decomposed as

$$T \wedge T = C \oplus Z$$

where C is the free $\mathbb{Z}_pP$-module of rank ℓ generated by $\{\kappa^i \wedge \kappa^{i-1} \mid i = 1, \ldots, \ell\}$, and Z is a free $\mathbb{Z}_p$-module generated by a certain $z \in T$ that is fixed by P. Thus, if $f \in \mathrm{Hom}_P(T \wedge T, T)$, then $f(z) = f(\theta z) = \theta f(z)$ for every $z \in Z$. Since the only fixed point in T under multiplication by θ is 0, this forces $f(z) = 0$. Thus, f is uniquely determined by its values on $\kappa^i \wedge \kappa^{i-1}$ for $i = 1, \ldots, \ell$. This motivates part a) of the following definition; for part b) recall that σ_i with i coprime to p denotes the Galois automorphism of $\mathbb{Q}_p(\theta)$ over $\mathbb{Q}_p$ defined by $\theta \mapsto \theta^i$.

Definition 9.4 Let $\delta_{i,j}$ be the Kronecker delta.

a) For $i \in \{1, \ldots, \ell\}$ the $\mathbb{Z}_pP$-homomorphism $\mathbb{T}_i : T \wedge T \to T$ is defined by

$$\mathbb{T}_i(\kappa^j \wedge \kappa^{j-1}) = \delta_{i,j} \quad \text{and} \quad \mathbb{T}_i(z) = 0. \tag{9.1}$$

b) For $a \in \{2, \ldots, \ell + 1\}$ the $\mathbb{Z}_pP$-homomorphism $\mathbb{S}_a : T \wedge T \to T$ is defined by

$$\mathbb{S}_a(x \wedge y) = \sigma_a(x)\sigma_{1-a}(y) - \sigma_a(y)\sigma_{1-a}(x). \tag{9.2}$$

c) For an ℓ-tuple $\mathbf{c} = (c_1, \ldots, c_\ell)$ of elements of $\mathbb{Q}_p(\theta)$ define

$$\mathbb{T}(\mathbf{c}) = c_1\mathbb{T}_1 + \ldots + c_\ell\mathbb{T}_\ell \quad \text{and} \quad \mathbb{S}(\mathbf{c}) = c_1\mathbb{S}_2 + \ldots + c_\ell\mathbb{S}_{\ell+1}.$$

An immediate consequence of Proposition 8.10 is that $\mathrm{Hom}_P(T\wedge T,T)$ is a free $\mathbb{Z}_p$-module with basis $\{\mathbb{T}_1,\ldots,\mathbb{T}_\ell\}$. One disadvantage of the generators $\mathbb{T}_i$ is that there is no closed formula for $\mathbb{T}_i(x\wedge y)$. This problem does not occur for the homomorphisms $\mathbb{S}_a$, however, as demonstrated below, a drawback of $\{\mathbb{S}_2,\ldots,\mathbb{S}_{\ell+1}\}$ is that this set does not generate $\mathrm{Hom}_P(T\wedge T,T)$ as a $\mathbb{Z}_pP$-module. The following was introduced by Leedham-Green & McKay [57, Section 8.3].

Proposition 9.5 *Let $f\in\mathrm{Hom}_P(T\wedge T,T)$.*

a) *There exists a unique $\mathbf{c}\in T^\ell$ with $f=\mathbb{T}(\mathbf{c})$.*

b) *There exists a unique $\mathbf{d}\in\mathbb{Q}_p(\theta)^\ell$ with $f=\mathbb{S}(\mathbf{d})$.*

c) *If $\mathbf{c}$ and $\mathbf{d}$ are as above, then $\mathbf{c}=\mathbf{d}B$ where B is the matrix*

$$B=\begin{pmatrix} b_{2,1} & b_{2,2} & \ldots & b_{2,\ell}\\ b_{3,1} & b_{3,2} & \ldots & b_{3,\ell}\\ \vdots & \vdots & \ddots & \vdots\\ b_{\ell+1,1} & b_{\ell+1,2} & \ldots & b_{\ell+1,\ell}\end{pmatrix}\in\mathbb{Q}_p(\theta)^{\ell\times\ell}. \tag{9.3}$$

with entries $b_{a,i}=(\theta^a-\theta^{a-1})\big((\theta^{1-a}-1)(\theta^a-1)\big)^{i-1}$.

Proof Part a) follows from Proposition 8.10 and the definition of $\mathbb{T}_1,\ldots\mathbb{T}_\ell$. We now consider the remaining parts. Note that $\mathbb{S}_a$ maps each $\kappa^i\wedge\kappa^{i-1}$ to $b_{a,i}$, and $\mathbb{S}_a=\mathbb{T}(\mathbf{b})$ for $\mathbf{b}=(b_{a,1},\ldots,b_{a,\ell})$ by part a). Thus, if $f=\mathbb{S}(\mathbf{d})$, then $f=\mathbb{T}(\mathbf{d}B)$ by the definition of B. We now show that B is invertible over $\mathbb{Q}_p(\theta)$. To this end, write $b_{a,i}=u_at_a^{i-1}$ were $u_a=\theta^a-\theta^{1-a}$ and $t_a=(\theta^{1-a}-1)(\theta^a-1)$, so

$$B=\begin{pmatrix} u_2 & & & \\ & u_3 & & \\ & & \ddots & \\ & & & u_{\ell+1}\end{pmatrix}\begin{pmatrix} 1 & t_2 & t_2^2 & \ldots & t_2^{\ell-1}\\ 1 & t_3 & t_3^2 & \ldots & t_3^{\ell-1}\\ & & & \vdots & \\ 1 & t_{\ell+1} & t_{\ell+1}^2 & \ldots & t_{\ell+1}^{\ell-1}\end{pmatrix}. \tag{9.4}$$

Using a Vandermonde determinant,

$$\det(B)=u_2\ldots u_{\ell+1}\prod_{2\leqslant a<b\leqslant\ell+1}(t_b-t_a).$$

Each $u_a=\theta^{1-a}(\theta^{2a-1}-1)\in T_1\backslash T_2$, and $t_b-t_a=\theta^{1-a}(\theta^{a-b}-1)(\theta^{a+b-1}-1)\in T_2\backslash T_3$ for all $2\leqslant a<b\leqslant\ell+1$. It follows that $\det(B)\in T_{\ell^2}\backslash T_{\ell^2+1}$; in particular, $\det(B)\neq 0$. This completes the proof. □

In Lemma 3.17 we used the valuation $\mathrm{val}\colon\mathbb{Q}_p(\theta)^\times\to\mathbb{Z}$ which maps a nonzero $s\in\mathbb{Q}_p(\theta)$ to the unique integer $z=\mathrm{val}(s)$ such that $s\in\mathfrak{p}^z\setminus\mathfrak{p}^{z+1}$, where $\mathfrak{p}^i=T_i$ is the ideal of $\mathbb{Z}_p[\theta]$ generated by κ^i. We extend this definition to ℓ-tuples $\mathbf{c}\in\mathbb{Q}_p(\theta)^\ell$ and say that

$$\mathrm{val}(\mathbf{c})=z\quad\text{if and only if}\quad\mathbf{c}\in(\mathfrak{p}^z)^\ell\setminus(\mathfrak{p}^{z+1})^\ell.$$

Lemma 9.6 *For $i = 1, \dots, \ell$ let $\mathbf{t}_i$ be the i-th row of B^{-1}, with B as in* (9.3).

a) *We have $\mathbb{T}_i = \mathbb{S}(\mathbf{t}_i)$ and* $\mathrm{val}(\mathbf{t}_i) = -2i + 1$.

b) *Let $f\colon T \wedge T \to T$ such that $f = \mathbb{S}(\mathbf{v})$ for some $\mathbf{v} = a_1\mathbf{t}_1 + \dots + a_\ell \mathbf{t}_\ell$ with each $a_i \in \mathbb{Q}_p(\theta)$. Then* $\mathrm{val}(\mathbf{v}) = \min\{\mathrm{val}(a_i\mathbf{t}_i) \mid i = 1, \dots, \ell\}$.

Proof a) Proposition 9.5 and its proof show that $\det(B) \in \mathfrak{p}^{\ell^2} \setminus \mathfrak{p}^{\ell^2+1}$, hence B is invertible over $\mathbb{Q}_p(\theta)$, and $\mathbb{T}_i = \mathbb{S}(\mathbf{t}_i)$. It remains to determine $\mathrm{val}(\mathbf{t}_i)$. By definition, each entry $b_{a,i}$ of B has valuation $\mathrm{val}(b_{a,i}) = 2i - 1$, so $B = UD$ where U is an $\ell \times \ell$ matrix with entries in the group of units $\mathcal{U}$ of $\mathbb{Z}_p[\theta]$, and $D = \mathrm{diag}(\kappa, \kappa^3, \dots, \kappa^{2\ell-1})$ is a diagonal matrix with the given entries. Since $\det(B) \in \mathfrak{p}^{\ell^2} \setminus \mathfrak{p}^{\ell^2+1}$, this implies that $\det(U) \in \mathcal{U}$. Cramer's rule for matrix inverses shows that each entry of U^{-1} lies in $\mathbb{Z}_p[\theta]$, and since $\det(U^{-1}) \in \mathcal{U}$, each row and column of U^{-1} contains at least one element of $T \setminus \mathfrak{p}$. In conclusion, the valuation of the i-th row of U^{-1} is 0, hence the valuation of the i-th row of $B^{-1} = D^{-1}U^{-1}$ is $-2i + 1$, as claimed.

b) Let $\mu = \min\{\mathrm{val}(a_i\mathbf{t}_i) \mid i = 1, \dots, \ell\}$ and $J = \{i \in 1, \dots, \ell \mid \mathrm{val}(a_i\mathbf{t}_i) = \mu\}$. Clearly, $\mathrm{val}(\mathbf{v}) \geqslant \mu$, and we show equality. If $|J| = 1$, then $\mathrm{val}(\mathbf{v}) = \mu$ follows readily; thus we suppose $n > 1$ and $J = \{i_1, \dots, i_n\}$ with $i_1 < \dots < i_n$. Note that $\mathrm{val}(\mathbf{v}) = \mu$ if and only if

$$\mathrm{val}(a_{i_1}\mathbf{t}_{i_1} + \dots + a_{i_n}\mathbf{t}_{i_n}) = \mu. \tag{9.5}$$

Since $\mathrm{val}(a_{i_j}\mathbf{t}_{i_j}) = \mu$ for all j, we can assume that $\mu = 0$: simply replace $\mathbf{v}$ by a multiple $\kappa^{-\mathrm{val}(\mathbf{v})}\mathbf{v}$. By part a), each $\mathrm{val}(\mathbf{t}_{i_j}) = -2i_j + 1$, so $\mathrm{val}(a_{i_j}\mathbf{t}_{i_j}) = 0$ implies $\mathrm{val}(a_{i_j}) = 2i_j - 1$, that is, there is a uniquely defined $c_{i_j} \in \{1, \dots, p-1\}$ such that

$$a_{i_j}\mathbf{t}_i \equiv c_{i_j}\kappa^{2i_j-1}\mathbf{t}_i \bmod (\mathfrak{p})^\ell.$$

Suppose, for a contradiction, that (9.5) is false, that is, $\mathrm{val}(a_{i_1}\mathbf{t}_{i_1} + \dots + a_{i_n}\mathbf{t}_{i_n}) > 0$. This means that $a_{i_1}\mathbf{t}_{i_1} + \dots + a_{i_n}\mathbf{t}_{i_n} \in (\mathfrak{p})^\ell$, so $c_{i_1}\kappa^{2i_1-1}\mathbf{t}_{i_1} + \dots + c_{i_n}\kappa^{2i_n-1}\mathbf{t}_{i_n} \in (\mathfrak{p})^\ell$. Thus, there exist $c_1, \dots, c_\ell \in \{0, \dots, p-1\}$, not all 0, such that

$$c_1\kappa^1\mathbf{t}_1 + c_2\kappa^3\mathbf{t}_2 + \dots + c_\ell\kappa^{2\ell-1}\mathbf{t}_\ell \in (\mathfrak{p})^\ell. \tag{9.6}$$

This implies that the rows of the matrix

$$M = \mathrm{diag}(\kappa, \kappa^3, \dots, \kappa^{2\ell-1})B^{-1} \in T^{\ell\times\ell}$$

are linearly dependent over the field $T/\mathfrak{p}$. Note that over $\mathbb{Q}_p(\theta)$

$$M^{-1} = B\mathrm{diag}(\kappa^{-1}, \kappa^{-3}, \dots, \kappa^{-2\ell+1}),$$

and the factorisation of B in (9.4) allows us to rewrite B as

$$B = \mathrm{diag}(x_2, \dots, x_{\ell+1})V(z_2, \dots, z_{\ell+1})\mathrm{diag}(\kappa^1, \kappa^3, \dots, \kappa^{2\ell-1})$$

where $x_a = \theta^{1-a}(1+\theta+\ldots+\theta^{2a-2})$ and $z_a = (1+\theta+\ldots+\theta^{a-1})(1+\theta+\ldots+\theta^{-a})$ for all a, and $V(z_2,\ldots,z_{\ell+1})$ is a Vandermonde matrix with parameters $z_2,\ldots,z_{\ell+1}$. This yields

$$M^{-1} = \mathrm{diag}(x_2,\ldots,x_{\ell+1})V(z_2,\ldots,z_{\ell+1}).$$

Note that each $x_a \equiv (2a-1) \bmod \mathfrak{p} \neq 0$, and all the entries $z_a \equiv (-a^2+a) \bmod \mathfrak{p} \neq 0$ are pairwise distinct. This shows that the determinant of M^{-1} is a unit in $T/\mathfrak{p}$, which contradicts Equation (9.6). Hence Equation (9.5) holds. □

9.2.2 The isomorphism problem for skeleton groups

We continue to denote by $S = P \ltimes T$ the infinite pro-p-group of coclass 1. Lemma 9.6 allows us to parametrise the surjective $\mathbb{Z}_p P$-homomorphisms $T \wedge T \to T_j$ by suitable ℓ-tuples over $\mathbb{Q}_p(\theta)$. Based on this, we now describe a convenient parametrisation of the skeleton groups in $\mathcal{G}(p,1)$; we achieve this through the following definitions.

We write $K = \mathbb{Q}_p(\theta)$ and for $n \in \mathbb{N}$ define

$$\Gamma_n = (\mathfrak{p}^n)^\ell B^{-1} \quad \text{and} \quad \Delta_n = \Gamma_n \setminus \Gamma_{n+1}. \tag{9.7}$$

By construction, for every $j \in \mathbb{N}_0$, the set Δ_j parametrises the set of all surjective $\mathbb{Z}_p P$-homomorphisms $T \wedge T \to T_j$, so

$$\{f \in \mathrm{Hom}_P(T \wedge T, T) \mid \mathrm{im}(f) = T_j\} = \{\mathbb{S}(\mathbf{d}) \mid \mathbf{d} \in \Delta_j\},$$

and $\mathbb{S}(\mathbf{c}) = \mathbb{S}(\mathbf{d})$ if and only if $\mathbf{c} = \mathbf{d}$. By Proposition 8.12, for every $\mathbf{d} \in \Delta_j$ and $m \in \{j,\ldots,2j-d\}$ we can define the skeleton group

$$G_{\mathbf{d},m} = G_{\mathbb{S}(\mathbf{d}),m}, \tag{9.8}$$

which has depth $m-j$ in the branch $\mathcal{B}_j$ of the coclass tree in $\mathcal{G}(p,1)$. Henceforth, we assume that $j \geqslant p$ and $m \in \{j,\ldots,2j-d\}$.

We now address the isomorphism problem: for $\mathbf{c},\mathbf{d} \in \Delta_j$, when is $G_{\mathbf{d},m} \cong G_{\mathbf{c},m}$? Proposition 8.24 shows that $G_{\mathbf{d},m} \cong G_{\mathbf{c},m}$ if and only if there exists an orbit isomorphism between the groups, and Proposition 8.19 shows that the latter holds if and only if there exists $\alpha \in \mathrm{Aut}(S)$ such that

$$\alpha(\,\mathbb{S}(\mathbf{c})(s \wedge t)\,) \equiv \mathbb{S}(\mathbf{d})(\alpha(s) \wedge \alpha(t)) \bmod T_m \quad (\text{for all } s,t \in T).$$

In fact, $\mathrm{Aut}(S)$ acts on the set of surjective $\mathbb{Z}_p P$-homomorphisms $T \wedge T \to T_j$: if $\beta \in \mathrm{Aut}(S)$ and $f\colon T \wedge T \to T_j$, then $\beta \cdot f$ is defined as

$$(\beta \cdot f)\colon T \wedge T \to T_j, \quad s \wedge t \mapsto \beta(f(\beta^{-1}(s) \wedge \beta^{-1}(t))). \tag{9.9}$$

The correspondence between the surjective homomorphisms in $\mathrm{Hom}_P(T \wedge T, T_j)$ and Δ_j allows us to define an $\mathrm{Aut}(S)$-action on Δ_j, namely, if $\mathbf{c} \in \Delta_j$, then $\alpha \cdot \mathbf{c} \in \Delta_j$

is the unique tuple that satisfies $\alpha \cdot \mathbb{S}(\mathbf{c}) = \mathbb{S}(\alpha \cdot \mathbf{c})$. These actions can be used to summarise the solution to the isomorphism problem as follows.

Proposition 9.7 *Let $j \geqslant p$, let $\mathbf{c}, \mathbf{d} \in \Delta_j$, and let $m \in \{j, \ldots, 2j - d\}$. Then the following is equivalent.*

a) *The skeleton group $G_{\mathbf{c},m}$ and $G_{\mathbf{d},m}$ are isomorphic.*

b) *There exists $\alpha \in \mathrm{Aut}(S)$ such that the image of $\alpha \cdot \mathbb{S}(\mathbf{c}) - \mathbb{S}(\mathbf{d})$ lies in T_m.*

c) *There exists $\alpha \in \mathrm{Aut}(S)$ such that $\alpha \cdot \mathbf{c} - \mathbf{d} \in \Gamma_m$.*

To apply this result, we need more information about $\mathrm{Aut}(S)$; in the following we write the elements of $S = P \ltimes T$ as tuples (θ^i, t) with $t \in T$ and $i \in \mathbb{Z}$. We denote by $\mathcal{G} = \mathrm{Gal}(\mathbb{Q}_p(\theta)/\mathbb{Q}_p)$ the Galois group of the extension $\mathbb{Q}_p(\theta)/\mathbb{Q}_p$. Proposition 3.14 shows that $\mathcal{G}$ is cyclic of order d, and consists of automorphisms σ_b defined by $\theta \mapsto \theta^b$ for $b \in \{1, \ldots, d\}$. We write $\mathcal{U} = \mathbb{Z}_p[\theta]^*$ for the unit group of $\mathbb{Z}_p[\theta]$; the structure of this group is discussed in Proposition 3.15.

Proposition 9.8 *Let $\pi\colon \mathrm{Aut}(S) \to \mathrm{Aut}(P) \times \mathrm{Aut}(T)$, $\alpha \mapsto (\alpha|_{S/T}, \alpha|_T)$, be the restriction homomorphism.*

a) *The image of π is isomorphic to $\mathcal{U} \rtimes \mathcal{G}$, and a preimage of $(u, \sigma_b) \in \mathcal{U} \rtimes \mathcal{G}$ under π is given by*
$$\alpha(u, b)\colon S \to S, \quad (\theta^i, t) \mapsto (\theta^{ib}, u\sigma_b(t)).$$

b) *The kernel of π is isomorphic to $\mathrm{Z}^1(P, T)$ and a free $\mathbb{Z}_p$-module of rank d generated by $\{\alpha(1), \alpha(\theta), \ldots, \alpha(\theta^{d-1})\}$, where for $s \in T$ we define*
$$\alpha(s)\colon S \to S, \quad (\theta^i, t) \mapsto (\theta^i, t + (1 + \theta + \ldots + \theta^{i-1})s).$$

Proof Since $T \leqslant S$ is characteristic, we establish the claim by an application of Proposition 5.14.

a) Write $C = \mathrm{Comp}(P, T)$ and recall that
$$C = \{(\alpha, \beta) \in \mathrm{Aut}(P) \times \mathrm{Aut}(T) \mid \forall g \in P, t \in T : \beta(t^g) = \beta(t)^{\alpha(g)}\}.$$

Since S is a split extension, $C = \mathrm{im}\,\pi$. The automorphisms of $P = \langle \theta \rangle$ are $\mathcal{G}$. By fixing the $\mathbb{Z}_p$-basis $\{1, \theta, \ldots, \theta^{d-1}\}$ of T, we identify $\mathrm{Aut}(T)$ with $\mathrm{GL}_d(\mathbb{Z}_p)$, and the generator θ of P acts by multiplication by the companion matrix $c \in \mathrm{Aut}(T)$ of $1 + \mathrm{x} + \ldots + \mathrm{x}^{d-1}$ on T, see the proof of Proposition 3.16. It follows that
$$C = \{(\sigma_i, M) \mid \sigma_i \in \mathcal{G},\ M \in N_{\mathrm{GL}_d(\mathbb{Z}_p)}(\langle c \rangle),\ c^M = c^i\}.$$

For a fixed $\sigma_i \in \mathcal{G}$, an automorphism M of $\mathrm{Aut}(T)$ with $c^M = c^i$ is the composition of $\sigma_i \in \mathrm{Aut}(T)$ with an element of $C_{\mathrm{GL}_d(\mathbb{Z}_p)}(c)$; thus, $\mathrm{im}\,\pi = C \cong C_{\mathrm{GL}_d(\mathbb{Z}_p)}(c) \rtimes \mathcal{G}$. It remains to prove $C_{\mathrm{GL}_d(\mathbb{Z}_p)}(c) \cong \mathcal{U}$. The structure of c allows us to show, via a direct calculation, that the entries of $M \in \mathbb{Z}_p^{d \times d}$ with $cM = Mc$ are already determined by the first row of M. Since the $\mathbb{Z}_p$-linear combination $a_0 c^0 + a_1 c^1 + \ldots + a_{d-1} c^{d-1}$ has first row $(a_0, \ldots, a_{d-1})$ and commutes with c, we deduce that $C_{\mathbb{Z}_p^{d \times d}}(c) =$

$\{a_0c^0 + a_1c^1 + \ldots + a_{d-1}c^{d-1} \mid a_1, \ldots, a_{d-1} \in \mathbb{Z}_p\}$ is the $\mathbb{Z}_p$-algebra generated by c. This algebra is isomorphic to $\mathbb{Z}_p[\theta]$ via the isomorphism induced by $c \mapsto \theta$. In particular,

$$C_{\mathrm{GL}_d(\mathbb{Z}_p)}(c) = C_{\mathbb{Z}_p^{d\times d}}(c) \cap \mathrm{GL}_d(\mathbb{Z}_p) \cong \mathcal{U}.$$

In conclusion, $\operatorname{im}\pi \cong \mathcal{U} \rtimes \mathcal{G}$, and the claim about the preimage follows from this construction.

b) The elements of $\ker\pi$ correspond to automorphisms $(\theta^i, t) \mapsto (\theta^i, t + \kappa(\theta^i))$ defined by $\kappa \in Z^1(P, T)$. The cocycle property implies that κ is uniquely defined by $\kappa(\theta)$, and if $\kappa(\theta) = s \in T$, then $\kappa(\theta^i) = (1 + \theta + \ldots + \theta^{i-1})s)$ for all i. Thus, the automorphism defined by κ is $\alpha(s)$. The claim follows since $T \to \ker\pi$, $u \mapsto \alpha(u)$, is $\mathbb{Z}_p$-linear. □

The automorphism $\alpha(s)$ of Proposition 9.8b) acts trivially on $\mathrm{Hom}_P(T \wedge T, T)$. It remains to investigate the action of $(u, \sigma_b) \in \mathcal{U} \rtimes \mathcal{G}$ via the automorphism $\alpha = \alpha(u, b)$ defined in Proposition 9.8a). How does α act on the homomorphism $f = \mathbb{S}(\mathbf{c})$? Recall that if $\mathbf{c} = (c_1, \ldots, c_\ell) \in \Delta_j$, then

$$f = c_1\mathbb{S}_2 + \ldots + c_\ell\mathbb{S}_{\ell+1}$$

as in Definition 9.4, and $\alpha \cdot f = (\alpha \cdot c_1\mathbb{S}_2) + \ldots + (\alpha \cdot c_\ell\mathbb{S}_{\ell+1})$. Note that $\alpha(t) = u\sigma_b(t)$ and $\alpha^{-1}(t) = (\sigma_b^{-1}(u))^{-1}\sigma_b^{-1}(t)$ for $t \in T$, and a short calculation using (9.9) shows that for every $a = 2, \ldots, \ell+1$ and $c \in \mathbb{Q}_p(\theta)$

$$\alpha \cdot c\mathbb{S}_a = u\sigma_a(u)^{-1}\sigma_{1-a}(u)^{-1}\sigma_b(c)\mathbb{S}_a.$$

We have seen the new factor in this coefficient before, namely, in (3.3) when we studied the homomorphism $\rho_a\colon \mathcal{U} \to \mathcal{U}$, $u \mapsto u^{-1}\sigma_a(u)\sigma_{1-a}(u)$. This allows us to denote the induced action of $(u, \sigma_b) \in \mathcal{U} \rtimes \mathcal{G}$ on $\mathbf{c} = (c_1, \ldots, c_\ell) \in \Delta_j$ by

$$(u, \sigma_b) \cdot \mathbf{c} = (\rho_2(u)^{-1}\sigma_b(c_1),\ \ldots,\ \rho_{\ell+1}(u)^{-1}\sigma_b(c_\ell)). \tag{9.10}$$

With this action, Proposition 9.7 can be rephrased as follows.

Proposition 9.9 *Let $j \geqslant p$, let $\mathbf{c}, \mathbf{d} \in \Delta_j$, and let $m \in \{j, \ldots, 2j - d\}$. Then $G_{\mathbf{c},m} \cong G_{\mathbf{d},m}$ if and only if there exists $(u, \sigma) \in \mathcal{U} \rtimes \mathcal{G}$ such that $(u, \sigma) \cdot \mathbf{c} - \mathbf{d} \in \Gamma_m$.*

Remarkably, this is a translation of the isomorphism problem for skeleton groups to an orbit problem that can be studied via p-adic number theory. Specifically, the group $\mathcal{U} \rtimes \mathcal{G}$ acts on the subset Δ_j of Γ_j modulo Γ_m. The results in Section 3.4 are useful to analyse this action in detail.

9.3 Galois trees in $\mathcal{G}(p, 1)$

As before, $S = P \ltimes T$ is the pro-p-group of coclass 1 where $T = (\mathbb{Z}_p[\theta], +)$ and $P = \langle\theta\rangle$. Let $\mathcal{T}$ be the associated maximal coclass tree. The detailed structure of

$\mathcal{T}$ is very complicated for $p \geqslant 7$, so we focus on subtrees such as the skeletons. The aim of this section is to restrict attention to *Galois trees*; these are subtrees of the skeletons that have particularly nice properties. The content of this section is from [6, 15]. In this section we consider $p \geqslant 7$. As before,

$$d = p - 1 \quad \text{and} \quad \ell = (p-3)/2.$$

Recall that $\mathcal{B}_n$ denotes the branch in $\mathcal{T}$ with root $S_n = S/\gamma_n(S)$, and $\mathcal{S}_n$ is the subtree of $\mathcal{B}_n$ consisting of all skeleton groups in $\mathcal{B}_n$; by definition, groups in $\mathcal{S}_n$ have depth at most $n - d$ in $\mathcal{B}_n$; we call $\mathcal{S}_n$ the n-th skeleton of $\mathcal{T}$.

Definition 9.10 Let G be a group in $\mathcal{T}$ and let $\mathcal{S}_n$ be a skeleton in $\mathcal{T}$.

a) The *Galois order* $\mathrm{gal}(G)$ of G is the p'-factor of $|\mathrm{Aut}(G)|$.
b) A *Galois tree* in $\mathcal{S}_n$ with Galois order h is a subtree $\mathcal{R}$ of $\mathcal{S}_n$ that is maximal with respect to the property that $\mathrm{gal}(G) = h$ for all G in $\mathcal{R}$.

By definition, each skeleton $\mathcal{S}_n$ is partitioned by its Galois trees. As a first step toward investigating Galois trees, we present the following results. Recall the definition of Γ_n and Δ_n from (9.7), the definition of the skeleton group $G_{\mathbf{x},m}$ from (9.8), and the action of $\mathcal{U} \rtimes \mathcal{G}$ on Δ_n from (9.10).

Theorem 9.11 *Let $p \geqslant 7$, let $n > d$, let $h \geqslant 1$, and let $e \in \{1, \ldots, n-d\}$. If $\mathbf{x}, \mathbf{y} \in \Delta_n$, then the following hold.*

a) *Every group at depth e in $\mathcal{S}_n$ is isomorphic to $G_{\mathbf{z},n+e}$ for some $\mathbf{z} \in \Delta_n$, and $G_{\mathbf{x},n+e} \cong G_{\mathbf{y},n+e}$ if and only if $(u,\varphi) \cdot \mathbf{x} \equiv \mathbf{y} \bmod \Gamma_{n+e}$ for some $(u,\varphi) \in \mathcal{U} \rtimes \mathcal{G}$.*
b) *The Galois order of $G_{\mathbf{x},n+e}$ is divisible by h if and only if $(u,\tau) \cdot \mathbf{x} \equiv \mathbf{x} \bmod \Gamma_{n+e}$ for some $u \in \mathcal{U}$ and $\tau \in \mathcal{G}$ of order h; in particular, h divides $|\mathcal{G}| = d$.*
c) *If G is a group in $\mathcal{S}_n$, then the Galois order of every descendant of G in $\mathcal{S}_n$ divides the Galois order of G.*

Proof a) This follows from the definition of $\mathcal{S}_n$ and Proposition 9.9.

b) The proof of Proposition 8.24 shows that an automorphism $G_{\mathbf{x},n+e} \to G_{\mathbf{x},n+e}$ of p'-order h is induced by some $\alpha \in \mathrm{Aut}(S)$ with $\alpha \cdot \mathbf{x} \equiv \mathbf{x} \bmod \Gamma_{n+e}$ such that $\alpha(P) = P$ and $\alpha(T) = T$. Proposition 9.8 now shows that $\alpha = \alpha(u,b)$ for some $u \in \mathcal{U}$ and $\tau = \sigma_b \in \mathcal{G}$, that is, $(u,\tau) \cdot \mathbf{x} \equiv \mathbf{x} \bmod \Gamma_{n+e}$. It remains to show that $|\tau| = h$. Suppose, for a contradiction, that $|\tau| = m < h$. Then $\alpha(u,b)^m = \alpha(v,1)$ for some $v \in \mathcal{U}$, and $\alpha(v,1)$ induces an automorphism of $G_{\mathbf{x},n+e}$ with p'-order $h/m > 1$. Recall from Proposition 3.15 that $\mathcal{U} = \langle \omega \rangle \times \mathcal{U}^{(1)}$ where ω is a primitive d-th root of unity and $\mathcal{U}^{(1)} = 1 + \mathfrak{p}$; in particular, finite quotients of $\mathcal{U}^{(1)}$ have p-power order. It follows that $v = \omega^i$ for some $i \in \{1, \ldots, d-1\}$. Then $(v,1) \cdot \mathbf{x} = \omega^{-i}\mathbf{x} \not\equiv \mathbf{x} \bmod \Gamma_{n+e}$, which is a contradiction to $(u,\tau) \cdot \mathbf{x} \equiv \mathbf{x} \bmod \Gamma_{n+e}$. Thus, $|\tau| = h$.

Conversely, if $(u,\tau) \cdot \mathbf{x} \equiv \mathbf{x} \bmod \Gamma_{n+e}$ for some $u \in \mathcal{U}$ and $\tau \in \mathcal{G}$ of order h, then $\alpha = \alpha(u,b)$ with $\sigma_b = \tau$ is an automorphism of S with $\alpha \cdot \mathbf{x} \equiv \mathbf{x} \bmod \Gamma_{n+e}$. Proposition 8.19 shows that α induces an automorphism of $G_{\mathbf{x},n+e}$ of order divisible by h. Thus, h divides the Galois order of $G_{\mathbf{x},n+e}$, as claimed.

c) Suppose the depth of G and H in $\mathcal{S}_n$ is e and f, respectively, with $0 \leqslant e < f$. Up to isomorphism, we can assume that $H = G_{\mathbf{z},n+f}$ and $G = G_{\mathbf{z},n+e}$ for some $z \in \Delta_n$. If H has Galois order h, then b) shows that there exists $(u, \tau) \in \mathcal{U} \rtimes \mathcal{G}$ with $|\tau| = h$ and $(u, \tau) \cdot \mathbf{z} \equiv \mathbf{z} \bmod \Gamma_{n+f}$. Then $(u, \tau) \cdot \mathbf{z} \equiv \mathbf{z} \bmod \Gamma_{n+e}$, and, as in the proof of b), Proposition 9.8 shows that there exists an automorphism of $\mathrm{Aut}(G)$ whose order is divisible by h. The claim follows. □

9.3.1 A local-to-global principle

Theorem 9.11 describes the groups of depth e in $\mathcal{S}_n$ whose Galois order is divisible by h via fixed points under some action modulo Γ_{n+e}; that is, they are fixed points in a *local* setting. Our next aim is to prove the following theorem, which shows that they can also be described via *global* fixed points.

Theorem 9.12 *Let $p \geqslant 7$ and $n \geqslant d$. Let G be a group of depth $e \geqslant 1$ in $\mathcal{S}_n$ and Galois order h. Then there exist $(v, \tau) \in \mathcal{U} \rtimes \mathcal{G}$, where τ has order h and $v \in \mathbb{Z}_p$ is a d-th root of unity, and a fixed point $\mathbf{x} \in \Delta_n$ of (v, τ) such that $G \cong G_{\mathbf{x},n+e}$.*

We prove this theorem using Theorem 9.17, but first establish some lemmas. Throughout this section, let $n \geqslant d$ and $e \in \{0, \ldots, n-d\}$. We fix a Galois order h and factorise $d = hk$. Denote by σ a generator of the Galois group $\mathcal{G}$, and let $\tau = \sigma^k$, so τ generates the subgroup of $\mathcal{G}$ of order h. In some results below we use the homomorphism

$$\chi\colon \mathcal{U} \to \mathcal{U}, \quad u \mapsto u\tau(u)^{-1},$$

and the decomposition

$$\mathcal{U} = \ker(\chi) \times \mathrm{im}(\chi)$$

proved in Proposition 3.20.

Lemma 9.13 *If G has depth e in $\mathcal{S}_n$, then $h \mid \mathrm{gal}(G)$ if and only if $G \cong G_{\mathbf{x},n+e}$ for some $\mathbf{x} \in \Delta_n$ such that $(v, \tau) \cdot \mathbf{x} \equiv \mathbf{x} \bmod \Gamma_{n+e}$ for some $v \in \ker(\chi)$.*

Proof Theorem 9.11 shows that $G \in \mathcal{S}_n$ and $h \mid \mathrm{gal}(G)$ if and only if $G \cong G_{\mathbf{y},n+e}$ for some $\mathbf{y} \in \Delta_n$ such that $(u, \tau) \cdot \mathbf{y} \equiv \mathbf{y} \bmod \Gamma_{n+e}$ for some $u \in \mathcal{U}$. Write $u = vw$ with $v \in \ker(\chi)$ and $w \in \mathrm{im}(\chi)$, that is, $w^{-1} = s\tau(s)^{-1}$ for some $s \in \mathcal{U}$. Now define $\mathbf{x} = (s, 1) \cdot \mathbf{y}$; Theorem 9.11 shows that $G \cong G_{\mathbf{x},n+e}$ and, by construction, modulo Γ_{n+e},

$$\begin{aligned} \mathbf{x} \equiv\ & (s, 1)(u, \tau) \cdot \mathbf{y} \equiv (s, 1)(u, \tau)(s, 1)^{-1} \cdot \mathbf{x} \\ \equiv\ & (su\tau(s)^{-1}, \tau) \cdot \mathbf{x} \equiv (v, \tau) \cdot \mathbf{x} \bmod \Gamma_{n+e}. \end{aligned}$$

The proves the claim. □

Next, we investigate the action of τ on Γ_n. Let $\mathbb{L} = \{a \in \mathbb{Q}_p(\theta) \mid \tau(a) = a\}$ be the fixed field in $\mathbb{Q}_p(\theta)$ under the action of τ. We define

$$\mathcal{O}_{\mathbb{L}} = \mathbb{Z}_p[\theta] \cap \mathbb{L} \quad \text{and} \quad \mathcal{U}_{\mathbb{L}} = \mathcal{U} \cap \mathbb{L} = \ker(\chi).$$

As before, we denote by ω a primitive d-th root of unity in $\mathcal{U}$. For $i \in \{0, \dots, d-1\}$, let

$$\Sigma_{n,i} = \{\mathbf{x} \in \Gamma_n \mid (1, \tau) \cdot \mathbf{x} = \omega^i \mathbf{x}\}$$

and

$$\Omega_{n,i} = \{\mathbf{x} \in \Delta_n \mid (1, \tau) \cdot \mathbf{x} = \omega^i \mathbf{x}\}.$$

Note that $\Omega_{n,i} = \Sigma_{n,i} \setminus \Sigma_{n+1,i}$ for all $i \in \{0, \dots, d-1\}$.

Lemma 9.14 *Let* $\mathcal{I} = \{kj \mid j \in \{0, \dots, h-1\}\}$.

a) *One can decompose*

$$\Gamma_n = \bigoplus\nolimits_{i \in \mathcal{I}} \Sigma_{n,i}.$$

b) *For each* $i \in \mathcal{I}$ *the set* $\Sigma_{n,i}$ *is an* $\mathcal{O}_{\mathbb{L}}$*-sublattice of* Γ_n *such that* $(u, 1) \cdot \mathbf{x} \in \Sigma_{n,i}$ *for each* $\mathbf{x} \in \Sigma_{n,i}$ *and* $u \in \mathcal{U}_{\mathbb{L}}$.

Proof a) Recall that τ is defined as $\tau = \sigma^k$ where $d = hk$. Lemma 3.17 shows that, considered as a $\mathbb{Q}_p$-linear map, $\tau \colon \mathbb{Q}_p(\theta) \to \mathbb{Q}_p(\theta)$ is diagonalisable with $h = d/k$ distinct eigenspaces of $\mathbb{Q}_p$-dimension k, each with eigenvalues $\{\omega^i \mid i \in \mathcal{I}\}$. It follows that τ is also diagonalisable in the induced action on $\mathbb{Q}_p(\theta)^\ell$. Let V_i be the eigenspace with eigenvalues ω^i for $i \in \mathcal{I}$. Clearly, each $\Sigma_{n,i} = V_i \cap \Gamma_n$, and so $\bigoplus_{i \in \mathcal{I}} \Sigma_{n,i} \subseteq \Gamma_n$. It remains to show the other inclusion.
For this purpose, consider $\mathbf{w} \in \Gamma_n$ and write $\mathbf{w} = \sum_{i \in \mathcal{I}} \mathbf{w}_i$ where each $\mathbf{w}_i$ lies in V_i. Let z be the number of nonzero summands $\mathbf{w}_i$ in $\mathbf{w}$. We prove by induction on z that each $\mathbf{w}_i \in \Sigma_{n,i}$; then $\Gamma_n = \bigoplus_{i \in \mathcal{I}} \Sigma_{n,i}$ is established.
If $z = 0$ or $z = 1$, then the claim is obviously true. Now suppose that $z > 1$ and choose j with $\mathbf{w}_j \neq 0$. Both $(1, \tau) \cdot \mathbf{w}$ and $(\omega^{-j}, 1) \cdot \mathbf{w} = \omega^j \mathbf{w}$ lie in Γ_n, hence also their difference $\mathbf{u} = (1, \tau) \cdot \mathbf{w} - \omega^j \mathbf{w}$. By construction, $\mathbf{u} = \sum_{i \in \mathcal{I}} \mathbf{u}_i$ where each $\mathbf{u}_i = (\omega^i - \omega^j)\mathbf{w}_i \in V_i$, and $\mathbf{u}$ has at most $z - 1$ nonzero summands $\mathbf{u}_i$. By the induction hypothesis, $\mathbf{u}_i \in \Sigma_{n,i}$ for each i. Since $\omega^i - \omega^j \in \mathbb{Z}_p^*$ for all $i \neq j$ in $\mathcal{I}$, it follows from $\mathbf{u}_i = (\omega^i - \omega^j)\mathbf{w}_i \in \Sigma_{n,i}$ that $\mathbf{w}_i \in \Sigma_{n,i}$ for all $i \neq j$ in $\mathcal{I}$. In particular, $\mathbf{w} - \mathbf{w}_j = \sum_{i \in \mathcal{I} \setminus \{j\}} \mathbf{w}_i \in \Gamma_n$, and so $\mathbf{w}_j = \mathbf{w} - (\mathbf{w} - \mathbf{w}_j) \in \Gamma_n$ as well. Now clearly $\mathbf{w}_j \in \Sigma_{n,j}$, which completes the proof.
b) It remains to show that each $\Sigma_{n,i}$ is invariant under the scalar action of $\mathcal{O}_{\mathbb{L}}$ and the action of $\mathcal{U}_{\mathbb{L}}$ defined in (9.10). Let $\mathbf{x} \in \Sigma_{n,i}$, $a \in \mathcal{O}_{\mathbb{L}}$, and $u \in \mathcal{U}_{\mathbb{L}}$. First, note that

$$(1, \tau) \cdot (a\mathbf{x}) = \tau(a)((1, \tau) \cdot \mathbf{x}) = a\omega^i \mathbf{x} = \omega^i (a\mathbf{x}),$$

so $a\mathbf{x} \in \Sigma_{n,i}$. Second, recall that $(u, 1) \cdot \mathbf{x} = (\rho_2(u)^{-1} x_1, \dots, \rho_{\ell+1}(u)^{-1} x_\ell)$. Since $u \in \mathcal{U}_{\mathbb{L}}$, each $\rho_j(u) \in \mathcal{U}_{\mathbb{L}}$, so

$$(1, \tau) \cdot ((u, 1) \cdot \mathbf{x}) = (u, 1) \cdot ((1, \tau) \cdot \mathbf{x}) = (u, 1) \cdot \omega^i \mathbf{x} = \omega^i ((u, 1) \cdot \mathbf{x}),$$

that is, $(u, 1) \cdot \mathbf{x}) \in \Sigma_{n,i}$, as claimed. □

Our next goal is to provide a number-theoretic condition for the existence of global fixed points. For this we first investigate the eigenvalues of $(1, \sigma)$ on Δ_0.

Lemma 9.15 *The eigenvalues of* $(1,\sigma)$ *on* Δ_0 *are* $\{\omega^{-2i+1} \mid i = 1,\dots,\ell\}$.

Proof The proof of Lemma 9.14 shows that the action of $(1,\sigma)$ on Γ_0 is diagonalisable. If $\mathbf{f} \in \Delta_0$ is an eigenvector of $(1,\sigma)$, then Lemmas 9.6 and 3.17 imply that the corresponding eigenvalue is $\omega^{\mathrm{val}(\mathbf{f})}$.

Recall that $\Gamma_1 = \kappa\Gamma_0$, and Γ_0 is the $\mathbb{Z}_p[\theta]$-span of $\mathbf{t}_1,\dots,\mathbf{t}_\ell$. Thus, modulo p, the eigenvalues of $(1,\sigma)$ on $\Delta_0 = \Gamma_0 \setminus \Gamma_1$ coincide with the eigenvalues on the elementary abelian quotient Γ_0/Γ_1 of rank ℓ; in particular, the number of different eigenvalues is at most ℓ. We show that it is exactly ℓ. To prove this, we use the notation of Lemma 9.6 and define an isomorphism

$$\psi\colon C_p^\ell \cong \Gamma_0/\Gamma_1, \quad (a_1,\dots,a_\ell) \mapsto a_1\mathbf{t}_1 + \dots + a_\ell\mathbf{t}_\ell + \Gamma_1;$$

here we identify C_p^ℓ with the group of ℓ-tuples over the integers $\{0,1,\dots,p-1\}$ modulo p. Recall that each $\mathbf{t}_j$ has valuation $\mathrm{val}(\mathbf{t}_j) = -2j+1$. In the following, we let $(1,\sigma)$ act on C_p^ℓ by using the induced action via ψ; this action is diagonalisable. Our claim is that for each $i \in \{1,\dots,\ell\}$ there is, up to scalar multiples, a unique eigenvector $(a_1,\dots,a_i,0,\dots,0) \in C_p^\ell$ of $(1,\sigma)$ with $a_i \neq 0$, and that this eigenvector has eigenvalue $\omega^{-2i+1} \bmod p$. We use induction to prove this claim; once proved, the lemma follows.

Base case: For the base case consider $i = \ell$. Since the action of $(1,\sigma)$ on C_p^ℓ is diagonalisable, there must be an eigenvector $a = (a_1,\dots,a_\ell) \in C_p^\ell$ with $a_\ell \neq 0$. This eigenvector in C_p^ℓ corresponds to an eigenvector $\mathbf{f} = a_1\mathbf{t}_1 + \dots + a_\ell\mathbf{t}_\ell + \mathbf{g} \in \Delta_0$ for some $\mathbf{g} \in \Gamma_1$. Recall that $\Gamma_1 = \kappa\Gamma_0 \subseteq (\mathfrak{p}^{-2\ell+2})^\ell$. Since $a_\ell \in \{1,\dots,d\}$, we observe that $\mathrm{val}(a_\ell\mathbf{t}_\ell) = \mathrm{val}(a_\ell) + \mathrm{val}(\mathbf{t}_\ell) = \mathrm{val}(\mathbf{t}_\ell) = -2\ell+1$. Together with $\mathrm{val}(\mathbf{g}) \geqslant -2\ell+2$, Lemma 9.6 implies that $\mathrm{val}(\mathbf{f}) = -2\ell+1$. Thus, the eigenvalue of $\mathbf{f}$ is $\omega^{-2\ell+1}$, and so the eigenvalue of a is $\omega^{-2\ell+1} \bmod p$.

For uniqueness, consider an eigenvector $a' = (a'_1,\dots,a'_\ell) \in C_p^\ell$ with $a'_\ell \neq 0$; as shown, the eigenvalue is $\omega^{-2\ell+1}$. Without loss of generality, we can assume that $a_\ell = a'_\ell$. Suppose, for a contradiction, that a and a' are linearly independent, so $b = a - a' = (b_1,\dots,b_{\ell-1},0) \in C_p^\ell$ is an eigenvector of $(1,\sigma)$ with eigenvalue $\omega^{-2\ell+1}$. This eigenvector corresponds to an eigenvector $\mathbf{u} = b_1\mathbf{t}_1 + \dots + b_{\ell-1}\mathbf{t}_{\ell-1} + \mathbf{h} \in \Delta_0$ for some $\mathbf{h} \in \Gamma_1$. The eigenvalue of $\mathbf{u}$ is $\omega^{\mathrm{val}(\mathbf{u})} = \omega^{-2\ell+1}$, so $\mathrm{val}(\mathbf{u}) \equiv -2\ell+1 \bmod d$. The structure of $\mathbf{u}$ implies that $\mathrm{val}(\mathbf{u}) \geqslant -2\ell+2$, so $\mathrm{val}(\mathbf{u}) \geqslant -2\ell+p > 0$. But then $\mathbf{u} \in \Gamma_1$ and $b = (0,\dots,0)$, which is a contradiction. This proves that there is, up to scalar multiples, a unique eigenvector of $(1,\sigma)$ in C_p^ℓ of the form $(a_1,\dots,a_\ell)$ with $a_\ell \neq 0$, and that the corresponding eigenvalue is $\omega^{-2\ell+1} \bmod p$.

Induction hypothesis: Our induction hypothesis is that for each j between $i+1$ and ℓ there is, up to scalar multiples, a unique eigenvector of $(1,\sigma)$ in C_p^ℓ of the form

$$v_j = (c_{j,1},\dots,c_{j,j},0,\dots,0) \in C_p^\ell$$

with $c_{j,j} \neq 0$, and that the corresponding eigenvalue is $\omega^{-2j+1} \bmod p$. Each v_j corresponds to an eigenvector $\mathbf{w}_j = c_{j,1}\mathbf{t}_1 + \dots + c_{j,j}\mathbf{t}_j + \mathbf{h}_j \in \Delta_0$ with eigenvalue

ω^{-2j+1} for some $\mathbf{h}_j \in \Gamma_1$, with

$$\mathrm{val}(\mathbf{h}_j) \geqslant \mathrm{val}(c_{j,1}\mathbf{t}_1 + \ldots + c_{j,j}\mathbf{t}_j) = \mathrm{val}(c_{j,j}\mathbf{t}_j) = -2j + 1.$$

For the induction step, we show that there exists an eigenvector of this form for $j = i$.

Existence of eigenvector: Since $(1, \sigma)$ is diagonalisable, the induction hypothesis implies that $\{v_{i+1}, \ldots, v_\ell\}$ can be extended to a basis of eigenvectors; by the uniqueness property, each of the additional i eigenvectors has m-th entry 0 for $m = i + 1, \ldots, \ell$. It follows that there must exist an eigenvector

$$a = (a_1, \ldots, a_i, 0, \ldots, 0) \in C_p^\ell$$

of $(1, \sigma)$ with $a_i \neq 0$. It comes from an eigenvector

$$\mathbf{f} = a_1\mathbf{t}_1 + \ldots + a_i\mathbf{t}_i + \mathbf{g} \in \Delta_0$$

for some $\mathbf{g} \in \Gamma_1$. We investigate this eigenvector in the following, determine its eigenvalue, and then show uniqueness.

Eigenvalue: We show that f has eigenvalue ω^{-2i+1}. Since $a_i \neq 0$, Lemma 9.6 yields $\mathrm{val}(a_1\mathbf{t}_1 + \ldots + a_i\mathbf{t}_i) = -2i + 1$. Write $\mathbf{g} = u_1\mathbf{t}_1 + \ldots + u_\ell\mathbf{t}_\ell$ with $u_1, \ldots, u_\ell \in \mathfrak{p}$, and let $r = \mathrm{val}(\mathbf{g})$. For each j, either $u_j = 0$, or $\mathrm{val}(u_j\mathbf{t}_j) \geqslant -2j + 2$. Thus, Lemma 9.6 implies that $r \geqslant -2\ell + 2$. We make a case distinction on r.

Case (1): If $r \geqslant -2i + 1$, then Lemma 9.6 shows that $\mathrm{val}(u_j\mathbf{t}_j) \geqslant -2i + 1$ for each j. Note that

$$\mathbf{f} = (a_1 + u_1)\mathbf{t}_1 + \ldots + (a_i + u_i)\mathbf{t}_i + u_{i+1}\mathbf{t}_{i+1} + \ldots + u_\ell\mathbf{t}_\ell$$

and $a_i \in \{1, \ldots, d\}$ and $u_i \in \mathfrak{p}$. Hence, $\mathrm{val}((a_i + u_i)\mathbf{t}_i) = -2i + 1$, and therefore $\mathrm{val}(\mathbf{f}) = -2i + 1$ by Lemma 9.6. Thus, the eigenvalue of $\mathbf{f}$ is ω^{-2i+1}, as claimed. We define $\mathbf{w}_i = \mathbf{f}$ and $v_i = a$ and $\mathbf{h}_i = \mathbf{g}$, which proves the induction step in this case.

Case (2): It remains to consider $-2\ell + 2 \leqslant r < -2i + 1$. In this case, the eigenvalue of $\mathbf{f}$ is ω^r; we show that this is not possible by modifying $\mathbf{g}$ by our known eigenvectors $\mathbf{w}_{i+1}, \ldots, \mathbf{w}_\ell$ until we obtain a contradiction. Let $s \in \{1, \ldots, \ell\}$ be minimal with $\mathrm{val}(u_s\mathbf{t}_s) = \mathrm{val}(\mathbf{g}) = r$; such an s exists by Lemma 9.6. It follows from $r = \mathrm{val}(u_s\mathbf{t}_s) = \mathrm{val}(u_s) + \mathrm{val}(\mathbf{t}_s) = \mathrm{val}(u_s) + (-2s + 1)$ that $u_s \in \mathfrak{p}^{r+2s-1} \setminus \mathfrak{p}^{r+2s}$. But $u_s \in \mathfrak{p}$, so $r + 2s - 1 > 0$. Together with our assumption that $r < -2i + 1$, this shows that $s \in \{i + 1, \ldots, \ell\}$. By the induction hypothesis, there is an eigenvector $\mathbf{w}_s = c_{s,1}\mathbf{t}_1 + \ldots + c_{s,s}\mathbf{t}_s + \mathbf{h}_s \in \Delta_0$ of $(1, \sigma)$ with eigenvalue ω^{-2s+1}; recall that $\mathrm{val}(\mathbf{h}_s) \geqslant -2s + 1$ and $c_{s,1}, \ldots, c_{s,s} \in \{0, \ldots, p - 1\}$ with $c_{s,s} \neq 0$.

Let $m \in \mathfrak{p}^{r+2s-1} \setminus \mathfrak{p}^{r+2s}$ be an eigenvector of σ with eigenvalue ω^{r+2s-1}; recall that $m \in \mathfrak{p}$ since $r + 2s - 1 > 0$. Now $\mathbf{f}' = \mathbf{f} - m\mathbf{w}_s$ is an eigenvector of $(1, \sigma)$ with eigenvalue ω^r. Since $m \in \mathfrak{p}$, both $\mathbf{f}$ and $\mathbf{f}'$ correspond to $a = (a_1, \ldots, a_i, 0, \ldots, 0)$. In particular, we can write $\mathbf{f}' = a_1\mathbf{t}_1 + \ldots + a_i\mathbf{t}_i + \mathbf{g}'$ where $\mathbf{g}' = (\mathbf{g} - m\mathbf{w}_s)$ has the form

$$\mathbf{g}' = (u_1 - mc_{s,1})\mathbf{t}_1 + \ldots + (u_s - mc_{s,s})\mathbf{t}_s + u_{s+1}\mathbf{t}_{s+1} + \ldots + u_\ell \mathbf{t}_\ell - m\mathbf{h}_s \in \Gamma_1.$$

Note that $m\mathbf{h}_s \in \Gamma_{r+2s}$. Since both $u_s, mc_{s,s} \in \mathfrak{p}^{r+2s-1} \setminus \mathfrak{p}^{r+2s}$, we can replace m by a suitable scalar multiple such that $u_s - mc_{s,s} \in \mathfrak{p}^{r+2s}$, and so $\mathrm{val}((u_s - mc_{s,s})\mathbf{t}_s) > r$. The minimality of s implies that if $j < s$, then $r < \mathrm{val}(u_j \mathbf{t}_j) = \mathrm{val}(u_j) + (-2j+1)$ and $u_j \in \mathfrak{p}^{r+2j}$; in particular, $\mathrm{val}(u_j - mc_{s,j}) \geqslant r + 2j > r$. In conclusion,

$$(*) \quad \mathrm{val}(u_j - mc_{s,j}) > r \quad \text{for all } j \leqslant s.$$

We now make a another case distinction.

Case (2a): If $\mathrm{val}(\mathbf{g}') \geqslant -2i+1$, then, as shown in Case (1), $\mathbf{f}'$ (and $\mathbf{f}$) has eigenvalue $\omega^{-2i+1} = \omega^r$, so $r \equiv -2i+1 \bmod d$. Since $r < -2i+1$, this forces $r \leqslant -2i+1-d$. But this is a contradiction to $-2\ell+2 \leqslant r$.

Case (2b): Thus, $\mathrm{val}(\mathbf{g}') < -2i+1$, and therefore $\omega^r = \omega^{\mathrm{val}(\mathbf{f})} = \omega^{\mathrm{val}(\mathbf{g}')}$, so $\mathrm{val}(\mathbf{g}') \equiv r \bmod d$. Lemma 9.6 and $(*)$ now show that

$$\mathrm{val}(\mathbf{g}') \geqslant \min\{r+d, \mathrm{val}(u_{s+1}\mathbf{t}_{s+1}), \ldots, \mathrm{val}(u_\ell \mathbf{t}_\ell)\} \geqslant r = \mathrm{val}(\mathbf{g}).$$

Since $\mathrm{val}(\mathbf{g}') \geqslant r + d \geqslant -2\ell + 2 + d$ contradicts $\mathrm{val}(\mathbf{g}') < -2i+1$, we deduce that $\mathrm{val}(\mathbf{g}') = \mathrm{val}(\mathbf{g}) = r$; in particular, $s' > s$ with $\mathrm{val}(u_{s'}\mathbf{t}_{s'}) = r$. This also shows that Case (2b) is only possible if $s < \ell$. In conclusion, we have shown that $\mathbf{f}' = a_1\mathbf{t}_1 + \ldots + a_i\mathbf{t}_i + \mathbf{g}'$ with $\mathbf{g}' \in \Gamma_1$, such that if we write $\mathbf{g}' = u_1'\mathbf{t}_1 + \ldots + u_\ell'\mathbf{t}_\ell$, then the minimal $s' \in \{1, \ldots, \ell\}$ with $\mathrm{val}(u_{s'}\mathbf{t}_{s'}) = r$ satisfies $s' > s$. We now iterate this argument until, eventually, we end up in Case (2a), which produces the final contradiction. This contradiction shows that Case (2) is not possible, thus $r \geqslant -2i+1$ and the claim follows from Case (1).

Uniqueness: Now consider a second eigenvector $a' = (a_1', \ldots, a_i', 0, \ldots, 0) \in C_p^\ell$ of $(1, \sigma)$ with $a_i' \neq 0$; as proved above, the eigenvalue of a' is $\omega^{-2i+1} \bmod p$. Suppose, for a contradiction, that a and a' are linearly independent. Replacing a' by a suitable scalar multiple, we can assume that $a_i = a_i'$, so $c = a - a' = (c_1, \ldots, c_j, 0, \ldots, 0)$, with $c_j \neq 0$ and $j < i$, is also an eigenvector of $(1, \sigma)$ with eigenvalue ω^{-2i+1}. But this our earlier proof that an eigenvector of this shape must have eigenvalue ω^{-2j+1}. Thus, a and a' must be linearly dependent.

In conclusion, we have proved that, up to scalar multiples, there is a unique eigenvector of $(1, \sigma)$ of the form $(a_1, \ldots, a_i, 0, \ldots, 0)$ with $a_i \neq 0$, and the corresponding eigenvalue is ω^{-2i+1}. This completes the induction step, and the proof of the lemma.□

The previous lemma allows us to determine when $\Omega_{n,i} \neq \emptyset$.

Lemma 9.16 *Recall that $\tau = \sigma^k$ where $d = hk$.*

a) *The eigenvalues of $(1, \tau)$ on Δ_0 are $\omega^{k(-2j+1)}$ for $j = 1, \ldots, \ell$.*

b) *$\Omega_{n,i} \neq \emptyset$ if and only if $(1, \tau)$ has an eigenvector in Δ_0 with eigenvalue ω^{i-nk}.*

c) *$\Omega_{n,i} \neq \emptyset$ if and only if $i \equiv k(n-2j+1) \bmod d$ for some $j = 1, \ldots, \ell$.*

Proof Let $\mathbf{x} \in \Delta_0$ be an eigenvector of $(1, \tau)$ with eigenvalue ω^{i-nk}. By Proposition 3.18, there is an eigenvector $a \in \mathfrak{p}^n \setminus \mathfrak{p}^{n+1}$ of σ with eigenvalue ω^n. Since $\tau = \sigma^k$, it follows that a is an eigenvector of τ with eigenvalue ω^{kn}. Hence, $a\mathbf{x} \in \Delta_n$ and

$$(1, \tau) \cdot (a\mathbf{x}) = \omega^{kn} a \omega^{i-kn} \mathbf{x} = \omega^i (a\mathbf{x}),$$

so $a\mathbf{x} \in \Omega_{n,i}$, as claimed.

Conversely, if $\mathbf{x} \in \Omega_{n,y}$ is an eigenvector of $(1, \tau)$ with eigenvalue ω^i, then choose an eigenvector $a \in \mathfrak{p}^{-n} \setminus \mathfrak{p}^{-n+1}$ of τ with eigenvalue ω^{-kn}, and observe that $a\mathbf{x} \in \Delta_0$ satisfies

$$(1, \tau) \cdot (a\mathbf{x}) = \omega^{-kn} a \omega^i x = \omega^{i-kn} ax,$$

so $a\mathbf{x}$ is an eigenvector of $(1, \tau)$ with eigenvalue ω^{i-kn}. Lemma 9.15 shows that the eigenvalues of $(1, \sigma)$ on Δ_0 are $\{\omega^{-2j+1} \mid j = 1, \ldots, \ell\}$, hence the remaining claim follows from $\tau = \sigma^k$. □

The following theorem summarises our 'local-to-global' principle for skeleton groups: every skeleton group at depth e in $\mathcal{S}_n$ can be defined as $G_{\mathbf{z},n+e}$ where $\mathbf{z} \in \Omega_{n,j}$ is some 'global fixed point' of (ω^{-j}, τ) for some j. We note that Theorem 9.12 is an immediate consequence of Theorem 9.17d).

Theorem 9.17 *Let $p \geqslant 7$ and $n \geqslant d$. If G has depth $e \geqslant 1$ in $\mathcal{S}_n$, then the following are equivalent; recall our convention that $d = hk$:*

a) *$G \in \mathcal{S}_n$ with $h \mid \mathrm{gal}(G)$;*

b) *$G \cong G_{\mathbf{x},n+e}$ for some $\mathbf{x} \in \Delta_n$ and $u \in \mathcal{U}$ with $(u, \tau) \cdot \mathbf{x} \equiv \mathbf{x} \bmod \Gamma_{n+e}$;*

c) *$G \cong G_{\mathbf{y},n+e}$ for some $\mathbf{y} \in \Delta_n$ and $v \in \ker(\chi)$ with $(v, \tau) \cdot \mathbf{y} \equiv \mathbf{y} \bmod \Gamma_{n+e}$;*

d) *$G \cong G_{\mathbf{z},n+e}$ for some $\mathbf{z} \in \Omega_{n,i}$ with $i \in \{k(n - 2j + 1) \bmod d \mid j = 1, \ldots, \ell\}$.*

Proof Parts a,b,c) are Theorem 9.11 and Lemma 9.13, and listed for completeness. If d) holds, then $(1, \tau) \cdot \mathbf{z} = \omega^i \mathbf{z} = (\omega^{-i}, 1) \cdot \mathbf{z}$, and so $\mathbf{z} \in \Omega_{n,i} \subseteq \Delta_n$ is fixed by (ω^i, τ) with $\omega^i \in \ker \chi$, proving c).

It remains to show that c) implies d). Suppose $G \cong G_{\mathbf{y},n+e}$ for some $\mathbf{y} \in \Delta_n$ such that $(v, \tau) \cdot \mathbf{y} \equiv \mathbf{y} \bmod \Gamma_{n+e}$ for some $v \in \ker(\chi)$. Lemma 9.14 implies that we can decompose $\mathbf{y} = \mathbf{y}_0 + \mathbf{y}_k + \ldots + \mathbf{y}_{(h-1)k}$ with each $\mathbf{y}_j \in \Sigma_{n,j}$. Write $v = \omega^i \theta^q u$ for some $i, q \in \mathbb{Z}$ and $u \in \mathcal{U}_2$, so that each $\rho_s(v) = \omega^i \rho_s(u)$ with $\rho_s(u) \in \mathcal{U}_2$. In the following let $j \in \{0, k, \ldots, (h - 1)k\}$. Since $v \in \ker(\chi)$, that is, $\tau(v) = v$,

$$(1, \tau)(v, 1) \cdot \mathbf{y}_j = (v, \tau) \cdot \mathbf{y}_j = (v, 1)(1, \tau) \cdot \mathbf{y}_j = \omega^j (v, 1) \cdot \mathbf{y}_j = \omega^{j-i} (u, 1) \cdot \mathbf{y}_j.$$

Note that $(v, 1) \cdot \mathbf{y}_j \in \Sigma_{n,j}$, so $(v, \tau) \cdot \mathbf{y}_j \in \Sigma_{n,j}$, and the assumption $\mathbf{y} - (v, \tau) \cdot \mathbf{y} \in \Gamma_{n+e}$ together with Lemma 9.14 implies that for every j

$$\mathbf{y}_j - (v, \tau) \cdot \mathbf{y}_j = \mathbf{y}_j - \omega^{j-i} (u, 1) \cdot \mathbf{y}_j \in \Gamma_{n+e},$$

that is,

$$(*) \quad (u, 1) \cdot \mathbf{y}_j \equiv \omega^{i-j} \mathbf{y}_j \bmod \Gamma_{n+e}.$$

This shows that $(u^d, 1) \cdot \mathbf{y}_j \equiv \mathbf{y}_j \bmod \Gamma_{n+e}$. On the other hand, the action of $(u, 1)$ on Γ_0/Γ_{n+e} has p-power order, so there exists some $m \in \mathbb{Z}$ such that $(u^d, 1)^m$ and $(u, 1)$ induce the same action. The former acts trivially on $\mathbf{y}_j$ while the latter acts by multiplication by ω^{i-j}. Thus, if $\omega^{i-j} \neq 1$, then $(*)$ forces $\mathbf{y}_j \in \Gamma_{n+e}$. In conclusion, $\mathbf{y} \equiv \mathbf{y}_i \bmod \Gamma_{n+e}$, and since $\mathbf{y} \in \Delta_n$ we deduce that $\mathbf{y}_i \in \Delta_n$, so $\mathbf{y}_i \in \Omega_{n,i}$. Thus, $G \cong G_{\mathbf{y},n+e} = G_{\mathbf{y}_i,n+e}$, and we can take $\mathbf{z} = \mathbf{y}_i$. The special form of i follows from Lemma 9.16. □

9.3.2 The isomorphism problem for groups in a Galois tree

An immediate consequence of Theorem 9.17 is the following.

Theorem 9.18 *Let $p \geqslant 7$ and $n \geqslant d$. Every leaf of a Galois tree in $\mathcal{S}_n$ has depth* $\mathrm{dep}(\mathcal{S}_n) = n - d$.

Proof Let $\mathcal{R}$ be a Galois tree in $\mathcal{S}_n$ with Galois order h. Recall that $\mathrm{dep}(\mathcal{S}_n) = n - d$ by definition. Let G be a group in $\mathcal{R}$ at depth e in $\mathcal{S}_n$. Theorem 9.12 shows that $G \cong G_{\mathbf{x},n+e}$ for some $\mathbf{x} \in \Delta_n$ with $(1, \tau) \cdot \mathbf{x} = \omega^i \mathbf{x}$ for some i. The group $H = G_{\mathbf{x},2n-d}$ is a descendant of G in $\mathcal{S}_n$, hence $\mathrm{gal}(H)$ divides $\mathrm{gal}(G)$ by Theorem 9.11. On the other hand, Theorem 9.17 shows that $\mathrm{gal}(G)$ divides $\mathrm{gal}(H)$. Thus, H is a leaf in $\mathcal{R}$, and every leaf of $\mathcal{R}$ has depth $n - d = \mathrm{dep}(\mathcal{S}_n)$. □

To describe the structure of Galois trees in more detail, we must solve the isomorphism problem of skeleton groups defined by 'global Galois fixed points', see Theorem 9.12. We start with a preliminary lemma; as before, $\tau = \sigma^k$ has order h where $d = p - 1 = hk$.

Lemma 9.19 *Let $\mathbf{x}, \mathbf{y} \in \Sigma_{n,i}$ such that $(u, 1) \cdot \mathbf{x} \equiv \mathbf{y} \bmod \Gamma_{n+e}$ for some $u \in \mathcal{U}$. If $u = vw$ with $v \in \ker(\chi)$ and $w \in \mathrm{im}(\chi)$, then $(v, 1) \cdot \mathbf{x} \equiv \mathbf{y} \bmod \Gamma_{n+e}$.*

Proof The assumptions imply that, modulo Γ_{n+e},

$$\begin{aligned}
\omega^i \mathbf{x} &\equiv (1, \tau) \cdot \mathbf{x} \equiv (1, \tau)(u^{-1}, 1) \cdot \mathbf{y} \equiv (\tau(u)^{-1}, 1)(1, \tau) \cdot \mathbf{y} \\
&\equiv \omega^i (\tau(u)^{-1}, 1) \cdot \mathbf{y} \equiv \omega^i (\chi(u), 1)(u^{-1}, 1) \cdot \mathbf{y} \\
&= \omega^i (\chi(u), 1) \cdot \mathbf{x} \bmod \Gamma_{n+e},
\end{aligned}$$

hence $(\chi(u), 1)$ fixes $\mathbf{x}$ modulo Γ_{n+e}. Note that $\chi(u) = \chi(w)$, so $(\chi(w), 1)$ also fixes $\mathbf{x}$ modulo Γ_{n+e}. Applying the same argument to $(\chi(w), 1) \cdot \mathbf{x} \equiv \mathbf{x} \bmod \Gamma_{n+e}$ allows us to deduce that $(\chi^2(w), 1)$ fixes $\mathbf{x}$ modulo Γ_{n+e}, and, inductively, the same holds for $(\chi^m(w), 1)$ for all $m \in \mathbb{N}$.

Recall that $\mathcal{U}^{(j)} = 1 + \mathfrak{p}^j$ for $j \geqslant 1$, and Proposition 3.15 implies that $(\mathcal{U}^{(j)})^{[p]} = \mathcal{U}^{(j+d)}$. Thus, we can choose j large enough such that $V = \mathcal{U}^{(j)}$ acts trivially on Γ_n/Γ_{n+e}. Similar to the decomposition of $\mathcal{U} = \ker(\chi) \times \mathrm{im}(\chi)$, we can decompose $V = \ker(\chi|_V) \times \mathrm{im}(\chi|_V)$, and χ induces an automorphism of $J = \mathrm{im}(\chi)/\mathrm{im}(\chi|_V)$, denoted ψ. The group J is finite since it is a quotient of $\mathrm{im}(\chi)/\mathrm{im}(\chi)^{[p^r]}$ for some $r \geqslant 1$ and the latter group is finite because it is finitely generated abelian with finite exponent. It follows that ψ has finite order, say $m \in \mathbb{N}$. This shows that $\psi^m(\hat{w}) = \hat{w}$ for the coset $\hat{w} = w\,\mathrm{im}(\chi|_V) \in J$, that is, $\chi^m(w) = ws$ for some $s \in V$. Since $(s, 1)$ and $(\chi^m(w), 1) = (ws, 1)$ fix $\mathbf{x}$ modulo Γ_{n+e}, it follows that $(w, 1)$ fixes $\mathbf{x}$ modulo Γ_{n+e}. Now $\mathbf{y} \equiv (u, 1) \cdot \mathbf{x} = (v, 1)(w, 1) \cdot \mathbf{x} \equiv (v, 1) \cdot \mathbf{x} \bmod \Gamma_{n+e}$ proves the claim.□

The following is a step towards solving the isomorphism problem for groups in a Galois tree.

Proposition 9.20 *Let* $\mathbf{x} \in \Omega_{n,i}$ *and* $\mathbf{y} \in \Omega_{n,j}$*. Then* $G_{\mathbf{x},n+e} \cong G_{\mathbf{y},n+e}$ *if and only if* $i = j$ *and* $(v, \sigma^m) \cdot \mathbf{x} \equiv \mathbf{y} \bmod \Gamma_{n+e}$ *for some* $v \in \ker(\chi)$ *and* $m \in \{0, \dots, k-1\}$.

Proof By Theorem 9.11, the groups are isomorphic if and only if there exists $(u, \varphi) \in \mathcal{U} \rtimes \mathcal{G}$ with $(u, \varphi) \cdot \mathbf{x} \equiv \mathbf{y} \bmod \Gamma_{n+e}$. Now observe that modulo Γ_{n+e}

$$\begin{aligned}\omega^j \mathbf{y} &\equiv (1, \tau) \cdot \mathbf{y} \equiv (1, \tau)(u, \varphi) \cdot \mathbf{x}\\ &\equiv \omega^i(\tau(u), \varphi) \cdot \mathbf{x} \equiv \omega^i(\chi(u)^{-1}, 1)(u, \varphi) \cdot \mathbf{x}\\ &\equiv \omega^i(\chi(u)^{-1}, 1) \cdot \mathbf{y},\end{aligned}$$

that is, $(\chi(u)^{-1}, 1) \cdot \mathbf{y} \equiv \omega^{j-i}\mathbf{y} \bmod \Gamma_{n+e}$. Since the image of χ is contained in $\langle\theta\rangle \times \mathcal{U}_2$, the action on Γ_0/Γ_{n+e} induced by $(\chi(u)^{-1}, 1)$ has p-power order. This forces $\omega^{j-i} = 1$, so $j = i$, as claimed.

Recall that $\tau = \sigma^k$ and write $\varphi = \sigma^r$ with $r = qk + m$ where $m \in \{0, \dots, k-1\}$. This shows that $(1, \varphi) \cdot \mathbf{x} = (1, \sigma^m)(1, \tau^q) \cdot \mathbf{x} = \omega^{qi}(1, \sigma^m) \cdot \mathbf{x}$, and so

$$\mathbf{y} \equiv (u, \varphi) \cdot \mathbf{x} \equiv \omega^{qi}(u, 1)(1, \sigma^m) \cdot \mathbf{x} \equiv (u\omega^{-qi}, \sigma^m) \cdot \mathbf{x} \bmod \Gamma_{n+e}.$$

Thus, without loss of generality, $q = 0$ and $\varphi = \sigma^m$ for some $m \in \{0, \dots, k-1\}$. Note that $(1, \sigma^m) \cdot \mathbf{x} \in \Sigma_{n,i}$, and $(u, 1)$ maps $(1, \sigma^m) \cdot \mathbf{x}$ to $\mathbf{y}$ modulo Γ_{n+e}. Writing $u = vw$ with $v \in \ker(\chi)$ and $w \in \mathrm{im}(\chi)$, it follows from Lemma 9.19 that $(v, 1)$ maps $(1, \sigma^m) \cdot \mathbf{x}$ to $\mathbf{y}$ modulo Γ_{n+e}. The claim follows. □

Corollary 9.21 *Let* $\mathcal{R}$ *be a Galois tree in* $\mathcal{S}_n$ *with Galois order* h*. There exists* $i \in \{k(n-2j+1) \bmod d \mid j \in \{1, \dots, \ell\}\}$ *such that the groups in* $\mathcal{R}$ *can be defined by elements of* $\Omega_{n,i}$.

Proof Let G be the root of $\mathcal{R}$ at depth e in $\mathcal{S}_n$. Let H be an m-step descendant of G in $\mathcal{R}$ for $m \geqslant 1$. By Theorem 9.12, up to isomorphism, we can define $G = G_{\mathbf{x},n+e}$ and $H = G_{\mathbf{y},n+e+m}$ for some $\mathbf{x} \in \Omega_{n,i_1}$ and $\mathbf{y} \in \Omega_{n,i_2}$. But H is a descendant of G, so $G_{\mathbf{x},n+e} \cong G_{\mathbf{y},n+e}$. Proposition 9.20 now shows that $i = i_1 = i_2$, and $i \in \{k(n-2j+1) \bmod d \mid j \in \{1, \dots, \ell\}\}$ by Lemma 9.16. □

9.3.3 Galois trees with Galois order d

More can be said for Galois trees with Galois order $d = p-1$. In the following, $\tau = \sigma$ has order d, that is, $h = d$ and $k = 1$. In this case, $\ker\chi = \{u \in \mathcal{U} \mid \sigma(u) = u\} = \mathbb{Z}_p^*$, and Proposition 9.20 states the following.

Proposition 9.22 *Let* $\mathbf{x} \in \Omega_{n,i}$ *and* $\mathbf{y} \in \Omega_{n,j}$. *Then* $G_{\mathbf{x},n+e} \cong G_{\mathbf{y},n+e}$ *if and only if* $i = j$ *and* $(v,1)\cdot\mathbf{x} \equiv \mathbf{y} \bmod \Gamma_{n+e}$ *for some* $v \in \mathbb{Z}_p^*$.

This significantly simplifies the isomorphism problem and allows us to prove the following structure result for Galois trees of Galois order d.

Theorem 9.23 *Let* $n \geqslant p \geqslant 7$.

a) *There are* ℓ *Galois trees in* $\mathcal{S}_n$ *of Galois order* d, *called* $\mathcal{R}_{n,1},\dots,\mathcal{R}_{n,\ell}$. *Each* $\mathcal{R}_{n,i}$ *has a root* $R_{n,i}$ *at depth* 1 *in* $\mathcal{S}_n$.
b) *Let* H *be a group in* $\mathcal{R}_{n,i}$ *for some* $i \in \{1,\dots,\ell\}$ *of depth* $e < n-d$ *in* $\mathcal{S}_n$. *Then* H *has* p *immediate descendants in* $\mathcal{R}_{n,i}$ *if and only if*

$$(e \bmod d) \in \{2,4,\dots,d-2\} \setminus \{d-2i\};$$

otherwise, H *has one immediate descendant in* $\mathcal{R}_{n,i}$.

Theorem 9.23 shows that each Galois tree in $\mathcal{S}_n$ with Galois order d starts at depth 1 in $\mathcal{S}_n$, and it has a well-understood branching structure which depends on the parameter i of the root of the subtree; see Figure 9.2 for an illustration.
The next corollary is an immediate consequence.

Corollary 9.24 *If* $p \geqslant 7$, *then the coclass tree in* $\mathcal{G}(p,1)$ *has unbounded width.*

We prove Theorem 9.23 using the following two lemmas. The first considers groups with Galois order d at depth 1.

Lemma 9.25 *Let* $n \geqslant p \geqslant 7$. *The skeleton* $\mathcal{S}_n$ *has exactly* ℓ *groups at depth* 1 *with Galois order* d; *these are obtained as* $R_{n,i} = G_{\mathbf{c}_i,n+1}$ *for* $i = 1,\dots,\ell$ *and some* $\mathbf{c}_i \in \Omega_{n,j_i}$ *with* $j_i = (n-2i+1 \bmod d)$.

Proof Proposition 9.14 shows that $\Gamma_n = \Sigma_{n,0}\oplus\ldots\oplus\Sigma_{n,d-1}$, and Lemma 9.16 shows that $\Omega_{n,j_i} \neq \emptyset$ if and only if $j_i = ((n-2i+1) \bmod d)$ for some $i = 1,\dots,\ell$. In particular, by Lemmas 9.15 and 9.16 the eigenspace of $(1,\sigma)$ in Γ_n/Γ_{n+1} with eigenvalue $\omega^{j_i} \bmod p$ is 1-dimensional and spanned by an element induced by some $\mathbf{c}_i \in \Omega_{n,j_i}$. Theorem 9.17 and Proposition 9.22 prove that, up to isomorphism, the groups in $\mathcal{S}_n$ at depth 1 with Galois order d are $G_{\mathbf{c}_1,n+1},\dots,G_{\mathbf{c}_\ell,n+1}$. □

Now we consider the remaining part of the skeleton.

Lemma 9.26 *Let* $n \geqslant p \geqslant 7$, *fix* $i \in \{1,\dots,\ell\}$, *and let* $\mathcal{R}_{n,i}$ *be the Galois tree in* $\mathcal{S}_n$ *as in Theorem 9.25. Let* H *be a group in* $\mathcal{R}_{n,i}$ *at depth* $e \in \{1,\dots,n-d-1\}$ *in* $\mathcal{S}_n$. *The group* H *has* p *immediate descendants in* $\mathcal{R}_{n,i}$ *if and only if* $e \bmod d$ *lies in* $\{2,4,\dots,d-2\} \setminus \{d-2i\}$; *otherwise,* H *has one immediate descendant in* $\mathcal{R}_{n,i}$.

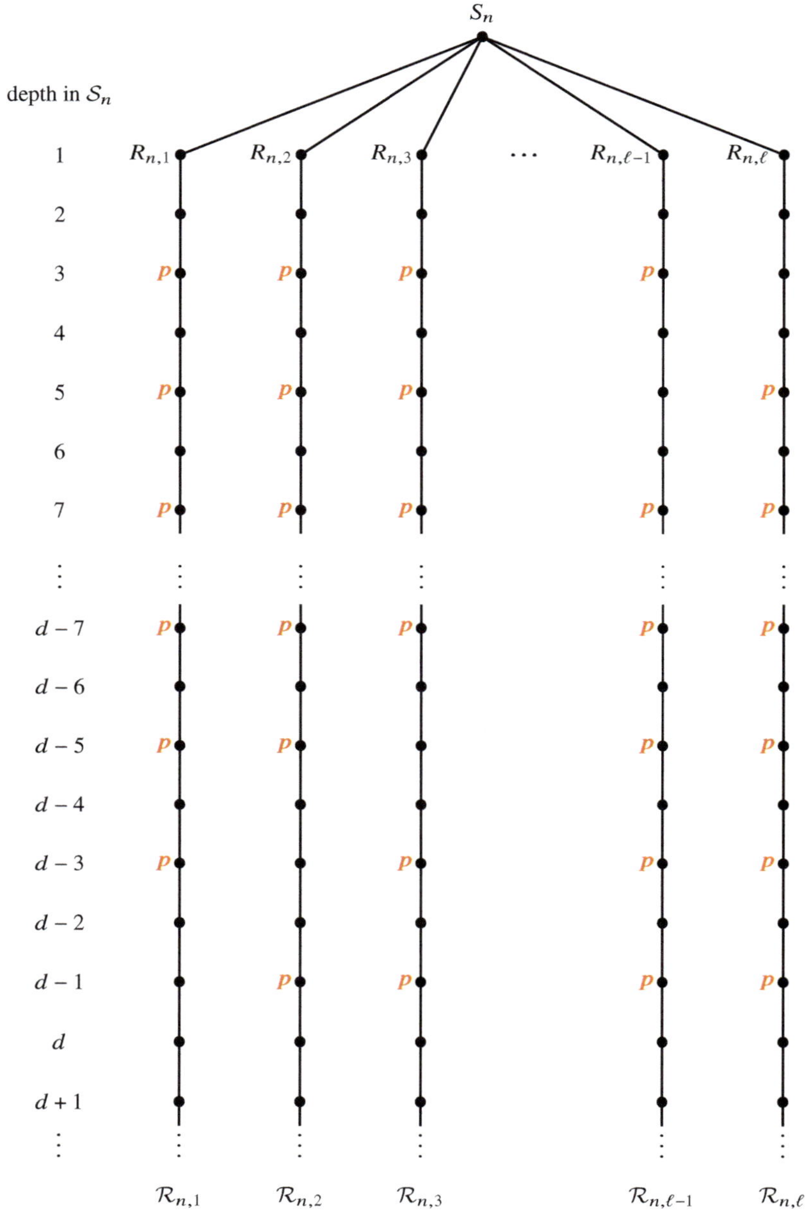

Fig. 9.2 The Galois trees $\mathcal{R}_{n,1}, \ldots, \mathcal{R}_{n,\ell}$ with roots $R_{n,1}, \ldots, R_{n,\ell}$ using Convention 2.11: an orange number p left of a vertex G means that the whole descendant tree $\Delta(G)$ of G appears p times such that the root of each of these descendant trees has the same immediate ancestor G.

Proof By Proposition 9.22 and Lemma 9.25, our aim is to determine the orbits and stabilisers of $\mathbb{Z}_p^*$ acting on

$$\Pi_{n,j,e} = \{\mathbf{c} + \Gamma_{n+e} \mid \mathbf{c} \in \Omega_{n,j}\},$$

where $j = j_i = (n - 2i + 1 \bmod d)$.

We first consider stabilisers; recall that $\mathbb{Z}_p^* = \langle\omega\rangle \times (1 + p\mathbb{Z}_p)$ and that $u \in \mathbb{Z}_p^*$ acts on $\mathbf{d} \in \Omega_{n,j}$ via

$$u \cdot \mathbf{d} = (u, 1) \cdot \mathbf{d} = u^{-1}\mathbf{d}.$$

It follows that $u \cdot \mathbf{d} \equiv \mathbf{d} \bmod \Gamma_{n+e}$ if and only if $u \in 1 + p^{\overline{e}}\mathbb{Z}_p$ with $\overline{e} = \lceil e/d\rceil$; note that $p^{\overline{e}}$ is the smallest p-power such that $p^{\overline{e}}\Gamma_n \leqslant \Gamma_{n+e}$. This implies that $\mathrm{Stab}_{\mathbb{Z}_p^*}(\mathbf{d}+\Gamma_{n+e}) = 1+p^{\overline{e}}\mathbb{Z}_p$. This stabiliser depends on e, but not on $\mathbf{d}$. In particular, by the Orbit-Stabiliser Theorem, each $\mathbb{Z}_p^*$-orbit has the same size $dp^{\overline{e}-1}$.

In the following, let $\mathcal{C} = \{0, 2, \dots, d-2\} \setminus \{d-2i\}$. We show that

$$|\Pi_{n,j,e}| = \begin{cases} |\Pi_{n,j,e+1}| & \text{if } e \bmod d \notin \mathcal{C} \\ p|\Pi_{n,j,e+1}| & \text{otherwise.} \end{cases}$$

To prove this, note that $|\Pi_{n,j,e}| < |\Pi_{n,j,e+1}|$ if and only if there exist $\mathbf{d}, \mathbf{e} \in \Omega_{n,j}$ with $\mathbf{d} \equiv \mathbf{e} \bmod \Gamma_{n+e}$ and $\mathbf{d} \not\equiv \mathbf{e} \bmod \Gamma_{n+e+1}$. This holds if and only if there exists $\mathbf{r} \in \Omega_{n+e,j}$ such that $\mathbf{d} = \mathbf{c} + \mathbf{r}$. By Lemma 9.16, such an $\mathbf{r}$ exists if and only if $j \equiv n + e - 2i' + 1 \bmod d$ for some $i' \in \{1, \dots, \ell\}$. As $j \equiv n - 2i + 1 \bmod d$, such an i' exists if and only if $e \equiv 2(i' - i) \bmod d$. A short calculation shows that

$$\mathcal{C} = \{2(i' - i) \bmod d \mid i' = 1, \dots, \ell\}.$$

In summary, $|\Pi_{n,j,e}| > |\Pi_{n,j,e+1}|$ if and only if $e \bmod d \in \mathcal{C}$, and in this case $|\Pi_{n,j,e}| = p|\Pi_{n,j,e+1}|$ since every eigenspace of $(1,\sigma)$ on $\Gamma_{n+e}/\Gamma_{n+e+1}$ has dimension 1.

We now reach the final conclusion. If $e \equiv 0 \bmod d$, then the size of the action domain grows by p, but also the size of the orbits grows by p, that is, the number of orbits remains stable; in this case, H has a single immediate descendant in $\mathcal{R}_{n,i}$. If $e \bmod d \in \mathcal{C} \setminus \{0\}$, then the size of the action domain grows by p, but the size of the orbits stays the same; in this case, H has p immediate descendants in $\mathcal{R}_{n,i}$. □

This completes the proof of Theorem 9.23.

Chapter 10
Towards Periodicity II

Coclass graphs are a useful tool to visualise the p-groups of fixed coclass. The Coclass Theorems imply that every coclass graph $\mathcal{G}(p, r)$ is a union of finitely many maximal coclass trees together with finitely many groups outside. Each coclass tree $\mathcal{T}$ is associated with an infinite pro-p-group S of coclass r. This pro-p-group S yields the mainline groups of $\mathcal{T}$ via nilpotent quotients $S/\gamma_j(S)$ and a significant part of the groups off the mainline via the skeleton group constructions.

The branches $\mathcal{B}_j$ of $\mathcal{T}$ satisfy Periodicity I. Computer experiments suggest that they also satisfy a second periodic pattern. Understanding this periodicity is currently one of the central goals of coclass theory. Achieving a complete understanding of it could facilitate a full classification of p-groups by coclass, which would represent a highly significant and fascinating result. This final chapter gives a brief overview of the state of the (currently open) Periodicity II.

10.1 The general case

We first consider coclass graphs $\mathcal{G}(p, r)$ for arbitrary p and r. Periodicity I shows that for every fixed integer $k \geqslant 1$ and large enough j, the sequence of pruned branches $\mathcal{B}_j(k), \mathcal{B}_{j+1}(k), \ldots$ is periodic; that is, $\mathcal{B}_i(k) \cong \mathcal{B}_{i+d}(k)$ for all $i \geqslant j$, where d is the dimension of the associated pro-p-group S.

Since every coclass tree in $\mathcal{G}(2, r)$ has bounded depth, this result shows that each graph $\mathcal{G}(2, r)$ can be described by a finite subgraph and one periodic pattern, proving the Coclass Graph Conjecture for $p = 2$. Moreover, the infinitely many finite 2-groups of coclass r can be partitioned into finitely many coclass families, each defined by a single parametrised group presentation. Theorem 7.3 asserts that the same is true for every pruned coclass graph. However, with the exception of $(p, r) = (3, 1)$, for odd primes the coclass graph $\mathcal{G}(p, r)$ contains trees of unbounded depth, and therefore Periodicity I is not capable to prove the Coclass Graph Conjecture.

The first evidence that supports a Periodicity II such as conjectured in Section 2.5 comprises extensive computations and theoretical results for $\mathcal{G}(5, 1)$ and $\mathcal{G}(3, 2)$. Based on these results, a variation of the periodicity depicted in Figure 2.5 was suggested by Eick, Leedham-Green, Newman & O'Brien [33, Conjecture W]. It

H. Dietrich and B. Eick, *Coclass graphs of p-groups*, Springer Monographs in Mathematics, https://doi.org/10.1007/978-3-032-17506-9_10

should be noted, though, that the graphs $\mathcal{G}(5,1)$ and $\mathcal{G}(3,2)$ possess a special property: all their coclass trees have bounded width, which makes their growth patterns considerably easier to analyse, both computationally and theoretically. In contrast, typical coclass trees have unbounded depth and width, presenting a far greater challenge.

10.2 The maximal class case

More is known in the special case of groups of coclass 1. Let $p \geqslant 5$ be a prime and $n \geqslant p+1$. If $\mathcal{B}_n$ is the n-th branch of the unique maximal coclass tree in $\mathcal{G}(p,1)$, then the depth of $\mathcal{B}_n$ lies between $e_p(n)$ and $e_p(n) + c_p$, where $e_p(n) = n - 2p + 8$ and $c_p = 4p - 19$ for $p \geqslant 7$, and $e_5(n) = n - 4$ and $c_5 = 4$. Dietrich [13, Theorem 1.1] proved for $n \geqslant p + 1$ that

$$\mathcal{B}_n(e_p(n)) \cong \mathcal{B}_{n+d}(e_p(n)).$$

Hence in the case of coclass 1, Periodicity I is not restricted to branches pruned at a fixed integer k, but also extends to branches pruned at positions determined by a function of p and n.

10.2.1 The impact of Galois orders

Dietrich [14] conjectured that for every non-terminal group G at depth $e_p(n)$ in $\mathcal{B}_{n+d}$ there is some group H at depth $e_p(n) - d$ in $\mathcal{B}_{n+d}$, called a *coclass parent* of G, such that the pruned descendant trees $\Delta(G)(d)$ and $\Delta(H)(d)$ are isomorphic. It remains open how to determine H for G. We recall the following result of [14].

Proposition 10.1 *Let $p = 5$ and $n \geqslant 6$. If G is a non-terminal group at depth $e_5(n)$ in $\mathcal{B}_{n+4}$ and H is a 4-step ancestor of G, then $\Delta(G)(4) \cong \Delta(H)(4)$.*

Thus, to prove the Coclass Graph Conjecture for $\mathcal{G}(5,1)$, it suffices to show that the sequence of subgraphs $\mathcal{B}_n \setminus \mathcal{B}_n(e_5(n))$ (comprising four layers of groups each) is periodic for $n \geqslant 10$.

Proposition 10.1 does not extend to larger primes as demonstrated by Saha [75, Theorem 6.43], who exhibited an infinite sequence of counterexamples in the graph $\mathcal{G}(7,1)$. New ideas for finding coclass parents are required; the following remark summarises some partial results. Recall that the Galois order of a p-group of maximal class is the p'-part of its automorphism group order, see Definition 9.10.

Remark 10.2 The Galois order of a group plays a role in determining periodic behaviour. More precisely:

a) For $p \equiv 5 \bmod 6$ and all large enough n, if G is a non-terminal group at depth $e_p(n)$ in $\mathcal{B}_{n+d}$ whose d-step ancestor H has Galois order 1, then $\Delta(G)(d)$ and

$\Delta(H)(d)$ are isomorphic, see [14, Theorem 1.2]. The restriction to $p \equiv 5 \bmod 6$ in the proof of [14, Theorem 1.2] is due to the observations made in Proposition 3.19.

b) A similar result holds if G has bounded distance from a *long* path in $\mathcal{B}_n$ on which the Galois orders of the groups do not change, see [14, Theorem 1.3] for the precise statement of this periodicity.

These results underpin the proof of Proposition 10.1 and have motivated the investigation of Galois trees; we also refer to the extended results by Cant, Dietrich, Eick & Moede [6], which concern Galois trees of order dividing $p-1$.

10.2.2 A new approach via the Lazard correspondence

A new approach to establishing Periodicity II for p-groups of maximal class was introduced by Eick, Komma & Saha [34]. Their work proposes a novel construction for the groups off the mainline, based on Lie p-rings and the Lazard correspondence. For background on the latter, we refer to Khukhro [49]. This approach extends existing ideas concerning skeleton groups and offers additional insight into a possible proof of Periodicity II. Whether it will ultimately lead to a complete proof of Periodicity II for groups of maximal class remains an open question.

References

1. N. Blackburn. On a special class of p-groups. Acta. Math. 100:45–92, 1958.
2. S. R. Blackburn, P. M. Neumann, and G. Venkataraman. Enumeration of finite groups. Cambridge University Press, 2007.
3. W. Bosma, J. Cannon, and C. Playoust. The MAGMA algebra system I: The user language. J. Symbolic Comput. 24:235–265, 1997.
4. K. S. Brown. Cohomology of Groups. Graduate Texts in Mathematics 87, Springer, 1987.
5. D. Burrell. On the number of groups of order 1024. Comm. Algebra 50:2408–2410, 2022.
6. A. Cant, H. Dietrich, B. Eick, and T. Moede. Galois trees in the graph of p-groups of maximal class. J. Algebra 604:429–450, 2022.
7. J. F. Carlson, L. Townsley, L. Valeri-Elizondo, and M. Zhang. Cohomology rings of finite groups. Algebra and Applications. Springer, 2003.
8. J. F. Carlson. Coclass and cohomology. J. Pure Appl. Algebra 200:251–266, 2005.
9. J. H. Conway, H. Dietrich, and E. A. O'Brien. Counting groups: gnus, moas, and other exotica. Math. Intelligencer 30(2):2-15, 2008.
10. M. Couson. On the character degrees and automorphism groups of finite p-groups by coclass. PhD Thesis, TU-Braunschweig, 2013.
11. M. Couson. Character degrees of finite p-groups by coclass. J. Algebra 418:91-109, 2014.
12. M. Couson. Corrigendum to "On the classification of prime-power groups by coclass". arXiv:1403.2091, 2014.
13. H. Dietrich. Periodic patterns in the graph of p-groups of maximal class. J. Group Theory 13:851–871, 2010.
14. H. Dietrich. A new pattern in the graph of p-groups of maximal class. Bull. Lond. Math. Soc. 42(6):1073–1088, 2010.
15. H. Dietrich and B. Eick. Finite p-groups of maximal class with 'large' automorphism groups. J. Group Theory 20(2):227–256, 2017.
16. H. Dietrich and B. Eick. The origins of coclass theory. Arch. Math. (Basel) 121(5-6):495–510, 2023.
17. H. Dietrich, B. Eick, and D. Feichtenschlager. Investigating p-groups by coclass with GAP. In Computational group theory and the theory of groups, volume 470 of Contemp. Math. 45–61. Amer. Math. Soc., Providence, RI, 2008.
18. H. Dietrich and S. Saha. A note on skeleton groups in coclass graphs. Internat. J. Algebra Comput. 29(1):127–146, 2019.
19. J. D. Dixon, M. P. F. du Sautoy, A. Mann, and D. Segal. Analytic pro-p-groups. 2nd edition, Cambridge University Press, 2003.
20. S. Donkin. Space groups and groups of prime-power order, VIII. Pro-p groups of finite coclass and p-adic Lie algebras. J. Algebra 111:316–342, 1987.
21. D. S. Dummit and R. M. Foote. Abstract algebra. John Wiley & Sons, 2004.

H. Dietrich and B. Eick, *Coclass graphs of p-groups*, Springer Monographs in Mathematics, https://doi.org/10.1007/978-3-032-17506-9

22. M. du Sautoy. Counting p-groups and nilpotent groups. Inst. Hautes Etudes Sci. Publ. Math. 92:63 – 112, 2001.
23. B. Eick. On the number of infinite branches in the graph of all p-groups of coclass r. J. Group Theory 8:687–700, 2005.
24. B. Eick. Determination of the uniserial space groups with a given coclass. J. London Math. Soc. 71:622–642, 2005.
25. B. Eick. Automorphism groups of 2-groups. J. Algebra 300:91–101, 2006.
26. B. Eick. Schur multiplicators of finite p-groups with fixed coclass. Israel J. Math. 166:157–166, 2008.
27. B. Eick. Schur multiplicators of infinite pro-p-groups with finite coclass. Israel J. Math. 166:147–157, 2008.
28. B. Eick, and D. Feichtenschlager. Computation of low-dimensional (co)homology groups for infinite sequences of p-groups with fixed coclass. Internat. J. Algebra Comput. 21:635–649, 2011.
29. B. Eick and D. J. Green. The Quillen categories of p-groups and coclass theory. Israel J. Math. 206:183–212, 2015.
30. B. Eick and D. J. Green. Cochain sequences and the Quillen category of a coclass family. J. Aust. Math. Soc. 102(2):185–204, 2017.
31. B. Eick, C. R. Leedham-Green, and E. A. O'Brien. Constructing automorphism groups of p-groups. Comm. Algebra 30:2271–2295, 2002.
32. B. Eick and C. R. Leedham-Green. On the classification of prime-power groups by coclass. Bull. Lond. Math. Soc. 40(2):274–288, 2008.
33. B. Eick, C. R. Leedham-Green, M. F. Newman, and E. A. O'Brien. On the classification of groups of prime-power order by coclass: the 3-groups of coclass 2. Internat. J. Algebra Comput. 23(5):1243–1288, 2013.
34. B. Eick, P. Komma, and S. Saha. The frame of the graph associated with the p-groups of maximal class. Submitted, 2025.
35. B. Eick, M. F. Newman, and E. A. O'Brien. The class-breadth conjecture revisited. J. Algebra 300:384–393, 2006.
36. B. Eick and E. A. O'Brien. AutPGrp – a refereed GAP package. `https://gap-packages.github.io/autpgrp`
37. W. Felsch, J. Neubüser, and W. Plesken. Space groups and groups of prime-power order. IV. Counterexamples to the class-breadth conjecture. J. London Math. Soc. 24:113–122, 1981.
38. G. A. Fernández-Alcober. The exact lower bound for the degree of commutativity of a p-group of maximal class. J. Algebra 174:523–530, 1995.
39. G. Gamble, W. Nickel, E. A. O'Brien, and M. Horn. ANUPQ – a refereed GAP package.`https://gap-packages.github.io/anupq`
40. The GAP Group. GAP – Groups, Algorithms, and Programming. `https://gap-system.org`, 2025.
41. O. Garaialde Ocaña. Cohomology of uniserial p-adic space groups with cyclic point group. J. Algebra 493:79–88, 2018.
42. F. Q. Gouvêa. p-adic numbers – an introduction. 2nd edition, Springer, 2003.
43. H. Hasse. Number theory. Reprint of the 1980 edition. Springer, 2002.
44. P. J. Hilton and U. Stammbach. A course in homological algebra. Graduate Texts in Mathematics 4, Springer New York, 1997.
45. K. Hoffman and R. Kunze. Linear Algebra. 2nd edition. Prentice-Hall, 1971.
46. D. F. Holt, B. Eick, and E. A. O'Brien. Handbook of computational group theory. Discrete Math. Appl., CRC Press, 2005.
47. I. M. Isaacs. Character Theory of Finite Groups. Dover, New York, 1976.
48. R. Joshi and S. Sarkar. On the Schur multiplier of finite p-groups of maximal class. J. of Group Theory 26(3):533–545, 2023.
49. E. I. Khukhro. p-Automorphisms of finite p-groups. London Mathematical Society Lecture Note Series 246, 1997
50. G. Karpilovsky. The Schur multiplier, volume 2 of London Mathematical Society Monographs. Oxford University Press, 1987.

51. C. R. Leedham-Green. Pro-p groups of finite coclass. J. London Math. Soc. 50:43–48, 1994.
52. C. R. Leedham-Green. The structure of finite p-groups. J. London Math. Soc. 50:49–67, 1994.
53. C. R. Leedham-Green and S. McKay. On p-groups of maximal class I. Quart. J. Math. Oxford Ser. 27(107):297–311, 1976.
54. C. R. Leedham-Green and S. McKay. On p-groups of maximal class II. Quart. J. Math. Oxford Ser. 29(114):175–186, 1978.
55. C. R. Leedham-Green and S. McKay. On p-groups of maximal class III. Quart. J. Math. Oxford Ser. 29(115):281–299, 1978.
56. C. R. Leedham-Green and S. McKay. On the classification of p-groups of maximal class. Quart. J. Math. Oxford Ser. 35(139):293–304, 1984.
57. C. R. Leedham-Green and S. McKay. The structure of groups of prime power order. London Mathematical Society Monographs. Oxford Science Publications, 2002.
58. C. R. Leedham-Green, S. McKay, and W. Plesken. Space group and group of prime-power order. V. A bound to the dimension of space groups with fixed coclass. Proc. London Math. Soc. 52(3):73–94, 1986.
59. C. R. Leedham-Green, P. M. Neumann, and J. Wiegold. The breadth and the class of a finite p-group. J. London Math. Soc. 1: 409–420, 1969.
60. C. R. Leedham-Green and M. F. Newman. Space groups and groups of prime-power order. I. Arch. Math. (Basel) 35(3):193–202, 1980.
61. K. Lux and H. Pahlings. Representations of Groups. Cambridge Studies in Advanced Mathematics 124. Cambridge University Press, 2010.
62. S. McKay. The precise bound to the coclass of space groups. J. London Math. Soc. 50:488–500, 1994.
63. R. J. Miech. The metabelian p-groups of maximal class II. Trans. Amer. Math. Soc. 272:465–474, 1982.
64. P. Moravec. Schur multipliers of finite p-groups of given coclass. Israel J. Math. 185:189–205, 2011.
65. J. Neukirch. Algebraische Zahlentheorie. Springer, 2007. English translation: Algebraic Number Theory, vol. 322 of Grundlehren der mathematischen Wissenschaften, translated by N. Schappacher, Springer, 2010.
66. M. F. Newman. Groups of prime-power order. Groups-Canberra 1989, Lecture notes math. 1456, Springer, 49-62, 1990.
67. M. F. Newman and E. A. O'Brien. Classifying 2-groups by coclass. Trans. Amer. Math. Soc. 351:131–169, 1999.
68. E. A. O'Brien. The p-group generation algorithm. J. Symbolic Comput. 9:677–698, 1990.
69. J. Opgenorth, W. Plesken, and T. Schulz. Crystallographic algorithms and tables. Acta Cryst. Sect. A 54:517–531, 1998.
70. L. Ribes and P. Zalesskii. Profinite groups. Second edition. Ergebnisse der Mathematik und ihrer Grenzgebiete. 3. Folge. A Series of Modern Surveys in Mathematics 40. Springer-Verlag, Berlin, 2010.
71. A. M. Robert. A course in p-adic analysis. Springer, 2000.
72. D. J. S. Robinson. A course in the theory of groups. Springer-Verlag, New York, Heidelberg, Berlin, 1982.
73. J. Rodney. 2-groups of almost maximal class. J. Austral. Math. Soc. Ser. A, 19:343–357, 1975.
74. J. Rodney. 2-groups of almost maximal class: corrigendum. J. Austral. Math. Soc. Ser. A 35:307, 1983.
75. S. Saha. Finite p-groups and coclass theory. Monash University PhD Thesis, see doi.org/10.26180/5ed98eb14da30, 2020.
76. A. Shalev. The structure of finite p-groups: Effective proofs of the coclass conjectures. Invent. Math. 115:315–345, 1994.
77. A. Shalev. Finite p-groups. NATO Adv. Sci. Inst. Ser. C Math. Phys. Sci. 471:401–450, 1995.
78. A. Shalev and E. I. Zel'manov. Pro-p groups of finite coclass. Math. Proc. Camb. Phil. Soc. 111:417–421, 1992.
79. M. Suzuki. Group Theory I. Springer, 1981.

80. P. Symonds. Rank, coclass, and cohomology, Int. Math. Res. Not. IMRN 22:17399–17412, 2021.
81. A. Vera-López, M. A. García-Sánchez, J. M. Arregi, and L. Ormaetxea. On the degree of commutativity of p-groups of maximal class. J. Pure Appl. Algebra 215:53–71, 2011.
82. A. Wiman. Über p-Gruppen mit maximaler Klasse. Acta Math. 88:317–346, 1952.

Index

ancestor, 14
immediate ancestor, 14
k-step ancestor, 14

branch, 14
breadth, 69

capable, 14
centraliser, 3
centre, 1
characteristic, 2
class, 2
Class-Breadth Conjecture, 70
coboundaries, 74
cobreadth, 70
cochain maps, 73
coclass, 6
Coclass Conjectures, 13
coclass family, 109
coclass graph, 14
coclass splitting, 85
coclass tree, 14
maximal, 14
pruned, 101
cocycles, 74
cohomology group, 74
cohomology map, 74
compatible pairs, 78
connecting homomorphism, 81
constructible group, 135
Correspondence Theorem, 3

degree of commutativity, 102, 137
depth
bounded, 28
derived length, 2
derived series, 1
derived subgroup, 1
descendant, 14
immediate descendant, 14
k-step descendant, 14
descendant tree, 14
doc(G), 137

exact sequence, 81
exponent, 74
exponent-p class, 26
extension, 21
central, 21
equivalent, 77
split, 21
exterior square, 118

Frattini subgroup, 4
fully invariant, 2

Galois order, 146
Galois tree, 146

hypercentre, 69

inverse limit, 8
inverse system, 8

λ-descendant, 26
lattice, 54
linear system
linear inverse system, 8
local field, 40
lower central series, 1
lower exponent-p central series, 26

mainline, 14
maximal class, 6

H. Dietrich and B. Eick, *Coclass graphs of p-groups*, Springer Monographs in Mathematics, https://doi.org/10.1007/978-3-032-17506-9

maximal subgroup, 3
module, 15

nilpotent, 2
non-generator, 4
non-terminal, 14
normaliser, 3

one-units, 39, 40
orbit isomorphisms, 129

p-adic absolute value, 36
p-adic integers, 8, 38
p-adic metric, 37
p-adic rationals, 37
p-adic units, 38
p-adic valuation, 36
p-cover, 21
p-group, 2
p-multiplier, 21
parent, 14
Periodicity I, 28, 104
Periodicity II, 33
pro-p-group, 8
pro-p-polycyclic series, 98
pro-finite group, 8

$\mathbb{Q}_p$-class, 53

rank, 5
residue class field, 40

Schur multiplier, 114
semidirect product, 15
settled, 102
skeleton group, 123, 127
soluble, 2
space group
  dimension, 65
  p-adic space group, 65
  point group, 65
  pre-space group, 69
  translation subgroup, 65
  uniserial space group, 65
strongly isomorphic, 78
subnormal, 4

tensor square, 117
terminal, 14
topological group, 12
two-step centraliser, 49

uniquely divisible, 74
uniserial p-subgroup, 50
uniserial action, 47, 48
upper central series, 1

valuation ring, 40

width
  bounded, 28
wreath product, 51

List of Symbols

$[g, h]$	commutator, page 1
$[G : U]$	index, page 2
$\mathrm{Aut}(G)$	automorphism group, page 5
$\mathrm{Aut}_M(E)$	subgroup of $\mathrm{Aut}(E)$ that leaves $M \leqslant E$ invariant, page 79
$\mathrm{Aut}_M(G^*)$	setwise stabiliser of M in $\mathrm{Aut}(G^*)$, page 23
$\mathrm{b}(G)$	breadth of G, page 69
$\mathcal{B}_i(k)$	pruned branch, page 101
$\mathcal{G}(p, r)$	coclass graph, page 14
$\mathrm{cb}(G)$	cobreadth of G, page 70
$\mathrm{cc}(G)$	coclass, page 6
$\mathrm{cl}(G)$	(nilpotency) class, page 2
$\mathrm{Comp}(G, M)$	group of compatible pairs, page 78
$\Delta(G)$	descendant tree, page 14
Δ_n	tuples defining certain surjective homomorphisms, page 143
δ_n	cohomology map, page 74
$\exp(G)$	exponent of the group, page 74
$\mathbb{F}_q$	finite field with q elements, page 59
$\mathrm{gal}(G)$	Galois order of a p-group, page 146
$\gamma(G)$	last non-trivial term of the lower central series, page 14
$\gamma_i(G)$	term of lower central series, page 1
Γ_n	tuples defining certain homomorphisms, page 143
$\mathrm{GL}_r(p)$	general linear group, page 5
$\lambda(G)$	last non-trivial term in the lower exponent-p central series, page 26
$\langle X \rangle$	subgroup generated by X, page 4
$\mathcal{O}$	valuation ring, page 40
$\mathcal{U}^{(n)}$	n-th one units, page 40
$\mathbb{S}_a$	$\mathbb{Q}_p P$-homomorphism $\mathbb{K} \wedge \mathbb{K} \to \mathbb{K}$, where $\mathbb{K} = \mathbb{Q}_p(\theta)$, page 120
$G_{\mathbf{x},m}$	skeleton group defined by a tuple $\mathbf{x}$, page 143
$\mathbb{N}$	positive integers, page 3
$\mathbb{N}_0$	non-negative integers, page 3
$\nu_p(a)$	p-adic valuation, page 36

H. Dietrich and B. Eick, *Coclass graphs of p-groups*, Springer Monographs in Mathematics, https://doi.org/10.1007/978-3-032-17506-9

$\Phi(G)$	Frattini subgroup, page 4
$\mathbb{Q}_p$	p-adic rationals, page 37
$\mathrm{rk}(G)$	rank, page 5
$\mathrm{val}(\mathbf{c})$	integer z with $\mathbf{c} \in (\mathfrak{p}^z)^\ell \setminus (\mathfrak{p}^{z+1})^\ell$, page 141
$\mathbb{Z}$	integers, page 3
$\mathbb{Z}_p$	p-adic integers, page 38
$\mathbb{Z}_p^*$	p-adic units, page 38
$\zeta_i(G)$	term of upper central series, page 1
$A \wr B$	wreath product, page 51
$B^n(G, M)$	group of n-th coboundaries, page 74
$C^n(G, M)$	cochain maps, page 73
$C_G(U)$	centraliser, page 3
C_n	cyclic group of order n, page 4
$C_{(p,s)}$	cyclic group generated by the companion matrix of the p^s-th cyclotomic polynomial, page 52
D_n	dihedral group of order n, page 6
d_s	integer $(p-1)p^{s-1}$, for a fixed prime p, page 53
E_γ	the groups extension defined by $\gamma \in Z^2(G, M)$, page 76
G'	derived subgroup, page 1
$G \ltimes M$	semidirect product, page 15
G^*	p-cover $G^* = F/R^*$ with $R^* = [F, R]R^{[p]}$ if $G = F/R$, page 21
g^h	conjugate, page 1
G^n	n-fold direct product of copies of G, page 4
$G^{(i)}$	term of derived series, page 1
$G^{[n]}$	subgroup generated by all n-th powers, page 4
$G_{P^*,f,m}$	skeleton group, page 127
$H^n(G, M)$	n-th cohomology group, page 74
$M(G)$	Schur multiplier of G, page 114
$N \trianglelefteq G$	normal subgroup, page 3
$N_G(U)$	normaliser, page 3
Q_n	quaternion group of order n, page 6
$S(p^r)$	semidirect product $P \ltimes T$ with $P \cong C_{p^r}$ and $T = (\mathbb{Z}_p^d, +)$, page 35
$S_{\mathbb{Z}}(p^r)$	semidirect product $P \ltimes T$ with $P \cong C_{p^r}$ and $T = (\mathbb{Z}^d, +)$, page 16
SD_n	semidihedral group of order n, page 6
$T \otimes T$	tensor square, page 117
$T \wedge T$	exterior square, page 117
T_α	a module T whose action is twisted by α, page 86
$U < G$	proper subgroup, page 1
$U \leqslant G$	subgroup, page 1
V_s	maximal uniserial subgroup, page 53
W_s	maximal uniserial subgroup, page 53
$Z(G)$	centre, page 1
$Z^n(G, M)$	group of n-th cocycles, page 74

GPSR Compliance

The European Union's (EU) General Product Safety Regulation (GPSR) is a set of rules that requires consumer products to be safe and our obligations to ensure this.

If you have any concerns about our products, you can contact us on ProductSafety@springernature.com

In case Publisher is established outside the EU, the EU authorized representative is:

Springer Nature Customer Service Center GmbH
Europaplatz 3
69115 Heidelberg, Germany

Batch number: 10732090

Printed by Printforce, the Netherlands